the sugar club cookbook

the sugar club

cookbook

peter gordon

acknowledgements

This book is dedicated to my whole family, the Gordons, and our partners. To Michael McGrath especially. Big thanks to Ashley Sumner and Vivienne Hayman, who started the SUGAR CLUB in the first place and to my agent Felicity Rubinstein who introduced me to Hodder & Stoughton. Also, to all the cooks in all the countries I have travelled through and worked with. Lastly, a huge thank you to my Gran, Molly Gordon, who set me on my way.

conversions

In New Zealand, we only use metric, so please convert as follows, when using 'cups' in my recipes: One cup=250ml 1 tbsp=15ml

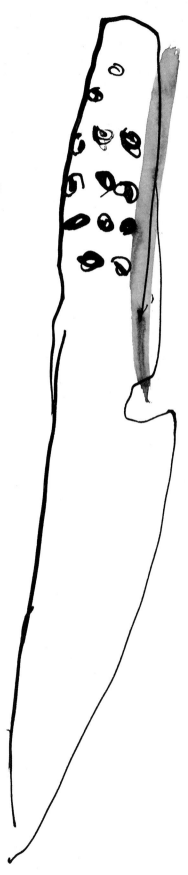

Inside design by
moira bogue
photographs & styling by
jean cazals
illustration by
trevor flynn

Peter Gordon would be happy to hear from readers with their comments on the book at the following e-mail address:
peter@foodware.demon.co.uk

contents

introduction

I spent many holidays and lived for periods as a child in New Zealand with my grand-mother, so it's only natural that I came to adopt her approach to food and its purpose:

sustenance, conversation & adventure.

Molly would send us kids traipsing off to the beach to collect seaweed. Some ended up in the soups or stews she made from cheap joints, while the rest slowly turned to mulch in the garden. She also had tubs of it fermenting away by the sheds behind the house, ready to dilute and feed to her pumpkins. These entwined themselves through the Macrocarpa trees – allowing her to pick a pumpkin without having to stoop. In the garden were all sorts of herbs, which she used to flavour, to heal, or both. The flowers of the comfrey went into salads, but when any of us had a sprain she made a poultice from its leaves – she knew, of course, that its popular name is 'Knit Bone'. For a boy growing up in New Zealand in the 60s and 70s it all seemed the norm. If Gran did it, then surely everyone else did too!

The things I learnt in my youth have stayed with me ever since: the generous use of fresh herbs and how a cheap cut of meat or the fleshless bone from a ham can transform a soup. My first cooking memory is helping my mother Timmy to cook an apple pie when I was about four: she did all the work while I just put the peeled and sugared pieces into the pie crust. My cousin Lynette introduced me to mushrooms – uncooked – marinated in olive oil and garlic. What a revelation! At home in Wanganui we never used vegetable oil to cook with. After Dad and his mates had cut up the carcasses of slaughtered beasts in the garage we'd turn the beef fat into soap by adding caustic soda and who knows what else. Almost all the food we ate was cooked in rendered fat. Scary.

I made my first scrapbook of recipes when I was five and then, at the age of seven, while cooking a dinner of fish and chips, I fell off my cooking stool and pulled the entire pot of boiling fat on to myself. Hospital, skin-grafts and more hospital followed. The family were convinced I'd lose my interest in food, but I didn't. After a short stint doing Horticultural Science at university I packed my bags for Melbourne, where I cooked and ate for five years. Melbourne – with its rich ethnic mix, numerous styles of food and a very high standard of living for most – must be one of the finest culinary centres in the world. As an apprentice I ate out once or twice a week, enjoying Vietnamese, Japanese, Greek, French and Thai cuisine. At the time Rogalsky's, Stephanie's, Berowra Waters

Top left: Vivienne and Ashley, owners of The Sugar Club. Top right: The Sugar Club, London. Photographs below: collage of Peter's family and travels, with Peter and Michael at centre.

and You and Me were all cooking 'New Australian' food, a reinvention of classical European cuisine. It was a minefield of ideas and a fantastic time. Once three friends and I drove 1000km to Sydney just to have lunch at Berowra Waters! I still remember the bone marrow with brioche and the baby chicken in salt crust with green sauce.

In 1985 I went to Bali and spent the next year travelling through Asia on my way to Europe. I was amazed by the food, religions, smells, colours, pace of life and, above all, by the brilliant people I met. Travelling through Indonesia, Malaysia, Singapore, Thailand, Burma, Nepal and India, I developed a great fondness for chillies, coconut, bamboo, spices, vegetables, relishes and anything warm and sunny. It was the most wonderful year of my life. After a brief spell in London I returned to Wellington, New Zealand, to set up the kitchen of the original Sugar Club. I'd been contacted in London by the owners, Vivienne Hayman and Ashley Sumner, who had heard of me through my cousins. They wanted to set up a place where one could go for interesting, eclectic and tasty food. Would I be interested in coming over and consulting for ten weeks to get it all going? The site they had was in the red-light district, in Vivian Street – an area known more for the transvestites and hookers who worked it than for diners. But the venture was a success and I had a great time – so much so that I stayed for two and a half years! One memorable night the street was cordoned off when the glue-sniffing kids squatting next door set fire to our building. The television crews outside were regular customers, as were most of the diners, so it was no problem. The glue sniffers' squat was eventually demolished and a garden centre now stands in its place. During my time at the Wellington Sugar Club I introduced our legendary beef pesto, various laksas, home-made goat's cheese, prawns in ginger and garlic, nori rolls, Bur Bur Cha Cha and much else besides.

In 1989 the Sugar Club was sold and I returned to London with the intention of opening another Sugar Club with Vivienne and Ashley, who were already scouting for suitable premises. The food in London restaurants was not what I was used to: it was either one style or another. There wasn't the pick and mix I had come to love from my travels and my childhood. By now I appreciated almost everything – from offal and green curries to buffalo yoghurt and deep-fried breads. As there didn't seem to be any places cooking eclectic food we were convinced we had a good idea and set about opening a place of our own.

We tried for nearly five years, but an assortment of obstacles prevented us from securing either a venue or the necessary finance. The closest we came was through discussions with a brewery, but they felt we were a risk as our views on food and service differed from theirs and the offer was withdrawn at the final meeting. We had worked on the deal for nearly a year and I found it very disheartening. I decided to pursue an independent career so that I could get into cooking without further delay. Ashley, Vivienne and I parted for two and a half years.

In January 1995 the site of the present Sugar Club was a run-down Asian restaurant on All Saints Road in Notting Hill. We were advised that the street was a no-go area, as had been said of the original site in Wellington. Still, we felt it was perfect: it was a good size, and it had a courtyard and lots of light. The planning and development were an exciting time. Where would the ovens go, and the bar and banquettes? After lots of hard work and effort, and much anxiety for Ashley and Vivienne, it was with a great feeling of achievement that the London incarnation of the Sugar Club finally opened its doors in June 1995. The first two weeks were a gradual build-up with friends – and friends of friends – filling the tables. At the end of the second week we received two excellent reviews and have never looked back since.

During the previous five years I'd worked at many restaurants and tried to do what I'm best at and feel comfortable with. I call this the 'magpie' approach to food – the mixing and matching of textures and tastes from the east to the west, sweet and sour, salty and spicy, soft and crunchy. If it tastes good it works.

My approach to food is that simple. It works or it doesn't. The careful mixing of sweet and savoury can really enhance a dish. Many people have difficulty appreciating the concept until they realise that chutneys are exactly that. As I have a mild allergy to dairy food I try to avoid it, but serve me a dessert without cream and I go to pieces. I love offal for the intensity of its texture and flavour. Chillies are tops – a judiciously placed bit of spiciness may go undetected yet will magically enhance a dish – even in desserts such as poached pears in red wine. I find it difficult to cook without an abundance of fresh herbs. Garlic, olive and sesame oils, tamari and Asian fish sauce are indispensable in my kitchen. Last year I discovered Spanish smoked paprika – both sweet and spicy. It transforms an aïoli, risotto, roast potatoes and so much else. Currently in the kitchen at the Sugar Club we have a string of dried aubergines hanging up – I'm not sure what they do with them in Turkey but no doubt they'll be on the menu before long! There is so much to learn about food and that's what is so exciting.

My style of food is loosely called 'Pacific Rim' or 'fusion', and it draws its influences from all over the world. It's simple to prepare and quick to make. I suggest you try it out and use the magpie technique to take from it what you want.

I would like to express my thanks to Simon Prosser, my editor, and Felicity Rubinstein, my agent. Felicity, who is a regular customer at the Sugar Club, thought that my cooking would be fairly easy to replicate in home kitchens – and so the idea for this book was born.

My thanks also go to everyone, from professional chefs I've worked with in kitchens from the Antipodes to London to cooks at roadside stalls from Bali to Bombay. Everyone I've worked with and everything I've eaten has influenced what I cook. Inspiration often comes from wanting to do something differently or better. So, especially to my fellow cooks and chefs, a big thank you!

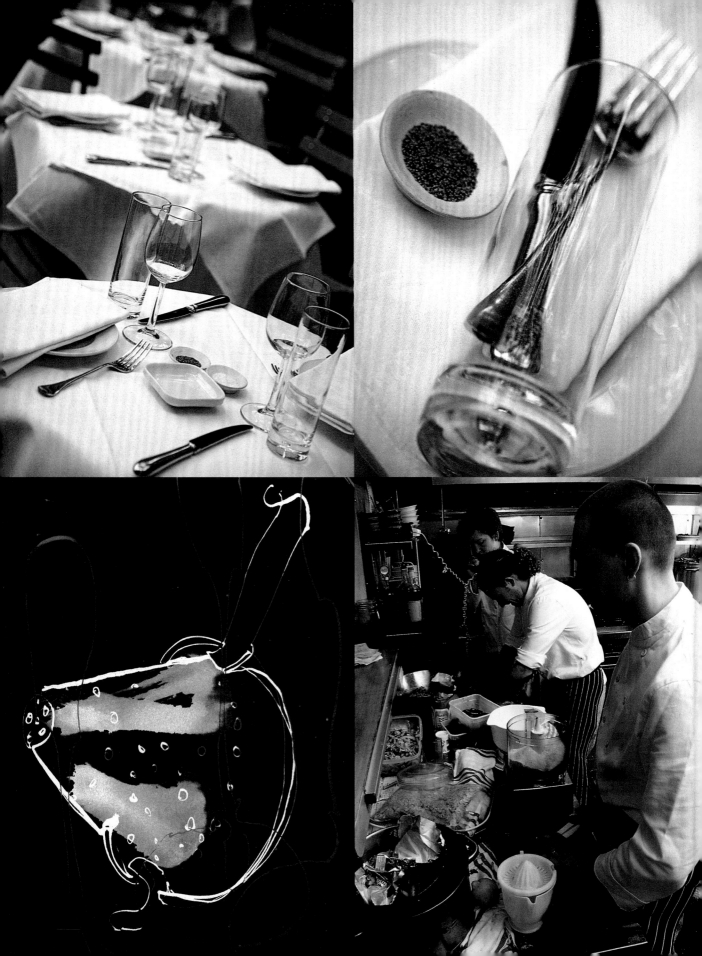

soups

spicy plantain,
tamarind & coconut soup
with coriander & mint

This is a really tasty soup that can also serve as a base for a meal. Sometimes at the Sugar Club we serve it with chicken or fish dumplings and lots of fresh coriander. You can also serve it with noodles, grilled fish pieces, shredded roasted duck or chunks of roasted pumpkin. You will need to use very ripe plantains to ensure the sweet taste that complements the sharpness of the tamarind. Look for fruit that is soft and has brown to black skin.

FOR SIX GENEROUS PORTIONS

100ml (3½fl oz) sesame oil
4 cloves of garlic, peeled and cut into quarters
1 green chilli, cut in half
2 teaspoons cumin seeds
1 teaspoon coriander seeds
2 red onions, peeled and sliced
2 very ripe plantains, peeled and cut into 1cm (⅓in) chunks
800ml (28fl oz) chicken or vegetable stock
600ml (1 pint) unsweetened coconut milk
250ml (9fl oz) tamarind paste
Tamari or sea salt
1 cup fresh coriander leaves
½ cup fresh mint leaves

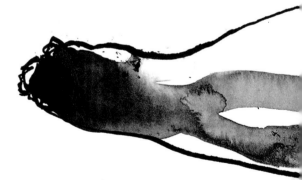

In a large pot, heat up the sesame oil and add the garlic, the chilli and the seeds; fry on a high heat for ½ minute, stirring well. Add the onions and fry on a high heat for another 2 minutes, stirring continuously to prevent it burning on the bottom. Add the plantain, stock, coconut milk and tamarind and bring to the boil. Turn the heat to a gentle simmer and put the lid on the pot. Cook until the plantain is soft, then purée. Taste for seasoning and add tamari or salt to taste. Stir in the coriander and mint just before serving.

spicy red lentil,
coriander & coconut soup
with chicken dumplings

FOR SIX LARGE BOWLS

2 teaspoons fennel seeds

2 teaspoons cumin seeds

2 teaspoons coriander seeds

1/4 cup sesame oil

1 leek, finely sliced and washed to remove grit

2 carrots, peeled and grated

2 red onions, peeled and finely sliced

2 lime leaves

4 cloves of garlic, peeled and roughly chopped

1 red chilli, finely sliced

1 thumb of fresh ginger, peeled and finely grated

300ml (11 fl oz) unsweetened coconut milk

1.4 litres (2 1/2 pints) water

3/4 cup red lentils

2 teaspoons Asian fish sauce

1/4 cup fresh lime juice

2 cups fresh coriander leaves

CHICKEN DUMPLINGS

2 chicken suprêmes

2 cloves of garlic, peeled

1/4 cup fresh coriander leaves

1/4 cup fresh mint leaves

1 tablespoon cornflour

2 teaspoons Asian fish sauce

To make the dumplings, remove any skin and bones from the chicken and cut each suprême into quarters. Put them in a food processor with the other five ingredients and process to a fine paste. Leave in the fridge for 30 minutes before rolling into grape-sized balls.

For the soup, lightly toast the seeds in a dry pan on the stove until they begin to smoke and smell nice, then let them cool before grinding finely in a spice mill. Sauté the next eight ingredients for 5 minutes over a moderate heat until a golden colour. Add the ground seeds, coconut milk and water and bring to the boil. Add the lentils and reduce the heat to a simmer, stirring occasionally to stop the lentils sticking to the bottom of the pan. Cook until the lentils are nearly done (about 20 minutes), then add the chicken dumplings and fish sauce. Boil gently until the dumplings are cooked (about 5 minutes) and taste for seasoning. Just before serving stir in the lime juice and coriander.

In many ways this soup is almost a laksa (see pages 23–29), but the lack of noodles and the addition of red lentils make it something a little different. It's delicious at any time of the year. In winter I make it thicker and heartier and in summer much hotter with chillies and serve it almost cool. The list of ingredients looks long, but it's fairly simple to make. Fish can be used in place of chicken, or use neither and make it a vegetarian dish.

sweet potato, leek, chickpea, parmesan & garlic soup

This is a great winter soup served thick and chunky with lots of crusty bread. If you want to serve it chilled in summer, purée well and thin a little with vegetable stock or water; like this it's deliciously refreshing. You can cook the chickpeas yourself, but if you use tinned it will save time and it doesn't make a noticeable difference in this recipe. The quantity of garlic may seem extravagant, but it's a great taste and a healthy one. Leave out the parmesan if you're on a dairy-free diet – the recipe will work just as well.

FOR SIX LARGE BOWLS

100ml (3 1/2 fl oz) olive oil
12 cloves of garlic, peeled and cut into quarters
3 leeks, sliced into 1cm (1/3 in) rings and washed well
12 fresh sage leaves, cut in half
3 bay leaves
1/2 lemon – remove the seeds and dice finely, skin and all
1 sweet potato, peeled and cut into 1cm (1/3 in) dice (about 2 cups)
2 cups cooked chickpeas, drained and rinsed
1 3/4 litres (3 pints) vegetable stock
Sea salt or Asian fish sauce
1 cup finely sliced spring onions
1/2 cup finely grated parmesan

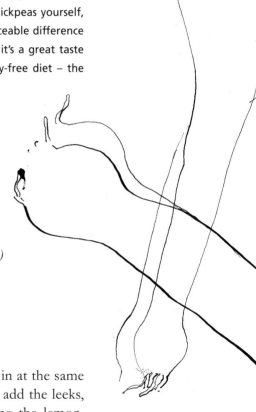

Heat the oil in a large pot and add the garlic – put it all in at the same time. Fry until the garlic just begins to colour and then add the leeks, sage and bay leaves. Sauté for 5 minutes before adding the lemon, sweet potato and chickpeas. Cover with stock and bring to the boil; the final thickness of the soup will depend on how much liquid you add at this point. Cook until the vegetables are tender and check for seasoning. At the last moment stir in the spring onions and parmesan.

spicy
carrot,
coconut & cumin soup

It's the sweetness of the carrots and the coconut that smooth out the spices in this soup. The turmeric imparts a subtle taste and colour, but here the cumin comes to the fore. Though best eaten on a cold day, this soup chills well and is an excellent picnic soup. The coriander or tarragon leaves stirred in at the time of serving add a fresh taste – either will complement the soup nicely.

FOR SIX BOWLS

1 kg (2 1/4 lb) carrots
6 red onions
4 cloves of garlic
1 3cm (1 in) thumb of ginger
1 medium-sized red chilli
50ml (1 3/4 fl oz) sesame oil
2 teaspoons cumin seeds
1/2 teaspoon fennel seeds
1 teaspoon turmeric powder
1 litre (1 3/4 pints) coconut milk
250ml (9 fl oz) water
Salt and cracked black pepper
1/2 cup fresh coriander leaves or 1/4 cup fresh tarragon leaves

Peel the carrots, onions, garlic and ginger and finely slice them and the chilli. Heat the oil almost to smoking and add the chilli, garlic, ginger, cumin and fennel. Fry on a fairly high heat, stirring constantly for 1 minute. Next add the onions and fry for a further 4 minutes, stirring occasionally, then add the carrots and turmeric and turn down the heat. Sauté for 10 minutes, stirring occasionally, then add the coconut milk and water and bring to the boil. Turn down and simmer for about 15 minutes until the carrots are soft. Remove from the heat and allow to cool before blending to a fine purée in a liquidiser. Test for seasoning and add salt and cracked black pepper.

Reheat just before serving and stir finely shredded coriander or tarragon leaves through it as it goes to the table.

chilled green soup
with olives

FOR SIX SERVINGS

1 leek
1 cup roughly chopped rocket leaves
2 cups roughly chopped watercress
2 cups roughly chopped curly parsley
1 cup pitted green olives
2 tablespoons rinsed capers
1 cup fresh basil leaves
1/2 cup fresh oregano leaves
1/2 cup fresh tarragon leaves
200ml (7fl oz) extra virgin olive oil
1 stalk celery
1 bunch spring onions
1 bunch chives
200ml (7fl oz) fresh lime or lemon juice
Salt and pepper

Slice the leek finely and wash it well. Bring 500ml (18fl oz) water to the boil in a pot and put in the leek, boil for 2 minutes, and then add the rocket, watercress and parsley. After 1 minute strain into a bowl, keeping the liquid, and then plunge the greens into a big bowl of cold water and leave to cool. Put the liquid in the fridge to chill.

Drain the greens again, blend them to a rough paste in a food processor and tip the paste into a bowl. Put the olives, capers, basil, oregano, tarragon and olive oil in the processor and blend to a rough paste; add this paste to the first. Finely dice the celery and slice the spring onions and chives, mix them into the paste and leave to sit for 30 minutes. Add the reserved liquid and the citrus juice plus another 300ml (11fl oz) of cold water and mix well, then test for seasoning and return to the fridge.

The soup is best left for a few hours to let the flavours develop; retaste for seasoning (as cold food needs more than hot) before serving in chilled bowls.

This is a delicious summer soup, but the colour deteriorates after a day so make only as much as you need. To keep the colour vivid the herbs are blended in oil before being exposed to the acidic citrus juice. This seals in their aroma and taste and protects the green colour against the acidity, which has to fight through a barrier of oil. It may sound a bit scientific, but it does help the soup!

butternut &
spaghetti squash soup
with toasted almonds & basil

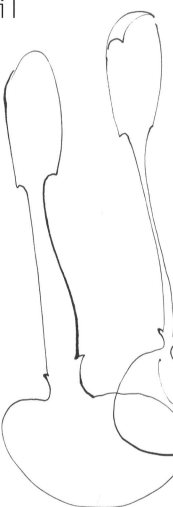

FOR SIX LARGE PORTIONS

$^1/_2$ spaghetti squash, cut lengthways

2 bay leaves

6cm (2$^1/_2$ in) sprig of fresh rosemary

1 white onion, peeled and cut into quarters

1 leek, cut into 1cm ($^1/_3$ in) rings and washed well

4 cloves of garlic, peeled

$^1/_2$ teaspoon fennel seeds

$^1/_2$ teaspoon coriander seeds

1 butternut squash, weighing around 750g (1lb 11oz),
    peeled, seeded and cut into 2cm ($^3/_4$ in) dice

1 cup lightly toasted flaked almonds

Salt or Asian fish sauce to season

1 cup lightly packed fresh basil leaves, cut into strips

50ml (1$^3/_4$ fl oz) extra virgin olive oil

Put the spaghetti squash into a large pot along with the next seven ingredients, cover with cold water and bring to the boil. Cook until you can scrape the flesh from the skin easily (it will lose its spaghetti shape if overcooked). When done, remove the squash and leave it to cool, then add the butternut squash to the pot and boil until cooked. When it is ready remove the bay leaves and rosemary stems and purée the butternut squash and liquid in a blender or food processor. Scrape the flesh from the spaghetti squash and gently break it up into strands. Mix with the puréed soup, then add the toasted almonds and reheat gently, tasting for seasoning. Serve in bowls, stirring in the basil and olive oil at the last minute.

Butternut squash is a delicious and versatile vegetable. If you have trouble finding good, fleshy pumpkins, you can happily use this squash instead. The spaghetti squash was something I only discovered in 1990 in London, where it is imported from Israel. It really does look like spaghetti once the flesh is removed from the skin. You can either boil them, as in this recipe, or cut them in half and roast on a greased tray – cut side down – until cooked, then scoop out the flesh.

parsnip, rosemary
& olive soup with feta

Hearty and rich, this soup is ideal for the winter. Thinned down a little, it works just as well in spring if you can still find good parsnips.

FOR FOUR BOWLS

80ml (3fl oz) olive oil
1 leek, finely sliced and washed well
400g (14oz) parsnips, peeled and cut into 1cm ($^1/_3$ in) dice
4 cloves of garlic, peeled and crushed
2 tablespoons fresh rosemary leaves, roughly chopped
800ml (28fl oz) vegetable stock
$^1/_2$ cup pitted olives, chopped
125g (4$^1/_2$ oz) feta cheese
Salt

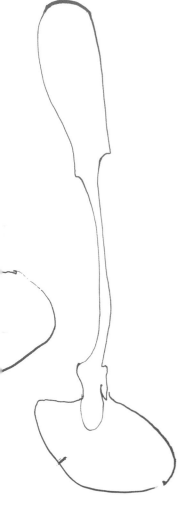

Heat the olive oil in a deep pot and sauté the leek for 5 minutes. Add the parsnips and turn the heat up a fraction, sauté until they begin to soften, then add the garlic and rosemary. Turn the heat up to full and fry for 1 minute, stirring continuously. Pour in the stock, bring to the boil, put the lid on and cook until the parsnips are done. Add the olives (use the best as cheap olives can be nasty) and cook for another minute. Serve in bowls and crumble in the feta. Salt may not be needed if the olives and feta came in strong brine.

laksas

crab & prawn
wonton laksa

This is a light and delicate laksa that makes an excellent first course or, if you bulk it out, a filling main course. Here the noodles are shredded 'wonton' wrappers – most good Asian food stores have them. You can also make the laksa from fish stock and use canned crab meat, which will save a lot of time. Otherwise, buy crabs that you know are meaty (look for specimens with big claws) as this will make the job of removing the flesh much easier. You may need a hammer or mallet to crack open the shells as some crabs are pretty tough critters. The dried shrimp paste is also readily available in Asian food stores – ask for 'kapi' or 'blachan'.

FOR EIGHT PORTIONS

1.5kg (3$^{1}/4$lb) crabs
2 medium-hot red chillies, cut in half
2 red onions, peeled and quartered
1 carrot, peeled and finely sliced
6 cloves of garlic, crushed with their skins on
3 teaspoons dried shrimp paste, roughly chopped
1 10cm (4in) thumb of fresh ginger, finely sliced
1 10cm (4in) thumb of galangal, finely sliced
400ml (14fl oz) chopped tomatoes (tinned are fine)
6 lime leaves
4 lemon-grass stems, bashed to flatten them
150g (5oz) raw prawn meat, finely minced
1 teaspoon fresh ginger, finely grated
Asian fish sauce
6 spring onions, finely sliced
150 wonton wrappers (this is an approximate number
  as you need enough to shred into noodles)
12 small bok choy, cut in half and washed
Lemon juice
1 cup fresh coriander leaves
$^{1}/4$ cup fresh mint leaves

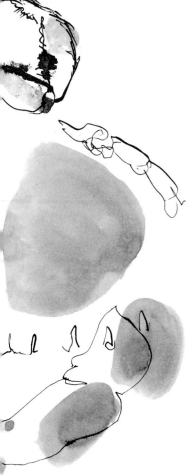

Preheat the oven to 220°C/425°F/Gas 7. Place the (humanely killed) crabs on a chopping board and lay a cloth on it. Using a hammer, or something similar, smash the body and the legs. Remove the large claws and crack them, but don't break them open. As you finish each crab put it and any juices that ooze out into a roasting dish; when they're all done place in the oven and cook for 30 minutes. When the 30 minutes are up take out the claws and add the chillies, onions, carrot, garlic and shrimp paste to the roasting dish. Cook for a further 20 minutes, then remove from the oven. Tip the contents into a large pot and cover with 3 litres (5 pints) of cold water; bring this to the boil, then turn down to a simmer. After 5 minutes skim off any scum, add the ginger, galangal, tomatoes, lime leaves and lemon grass and keep gently cooking for 2 hours. While it's cooking remove the meat from the claws by cracking them open gently. Put the meat aside and add the shells to the pot.

To make the filling for the wontons, mix the prawn meat, grated ginger, ½ teaspoon fish sauce and spring onions together. Put in the fridge to chill for 30 minutes. Take a wonton wrapper and drop 1 small teaspoon of the filling into the middle. Brush around the filling with water, or egg wash, and fold the edges together to form a tight bundle. Repeat until the filling is used up. Store on a tray covered with non-stick baking parchment or clingfilm.

Now strain the stock well as it will contain many small pieces of shell. This should give you about 2.5 litres (4½ pints) of crab broth. Return it to the boil, then turn to a rapid simmer and add the bok choy. Drop the wontons in one by one and cook for 4 minutes from the time you add the last. While the wontons are cooking, shred the remaining stack of wrappers into 5mm (⅕ in) 'noodles' and drop these in, dispersing them as you go so they don't stick together; cook for a further 30 seconds and turn off the heat. Test for seasoning and add more fish sauce and lemon juice to taste.

Divide the crab meat between the preheated bowls and scatter the coriander and mint over it. Then carefully ladle the soup, wontons, bok choy and noodles on top.

coconut, oyster & salmon
laksa

This laksa is really sumptuous. It can be made with lots of chilli because there's plenty of coconut milk to mask their heat. I recommend vermicelli rice noodles, but any type of noodle can be used. If you don't like biting into a warm, uncooked oyster, add it a little earlier in the cooking process. The roots of fresh coriander pack a lot of flavour, so if you're fortunate enough to be able to buy coriander whole wash the roots well and use them.

FOR SIX LARGE BOWLS

2 medium-hot red chillies (more if you like!)
4 cloves of garlic, peeled
1 6cm (2$^{1}/_{2}$in) thumb of ginger, peeled and roughly chopped
1 teaspoon finely ground coriander seeds
$^{1}/_{2}$ cup fresh coriander, roots, stems and leaves, all well washed
50ml (1$^{3}/_{4}$fl oz) sesame oil
250g (9oz) salmon fillet, skinned and bones removed, sliced into 12 pieces
50ml (1$^{3}/_{4}$fl oz) fresh lemon juice
1.2 litres (2 pints) unsweetened coconut milk
800ml (28fl oz) fish or vegetable stock
50ml (1$^{3}/_{4}$fl oz) Asian fish sauce
12 freshly opened oysters (save the juices that come out when you open them)
200g (7oz) dried vermicelli noodles, cooked as described on the packet
18 fresh mint leaves
3 spring onions, finely sliced

Put the first six ingredients into a food processor and purée to a coarse paste. Mix the salmon and lemon juice together and leave to marinate at room temperature while you make the laksa. Heat a large pot and add the paste; fry for 1 minute, stirring well. Add the coconut milk and stock and bring to the boil. Simmer for 10 minutes, then add the fish sauce, the oysters and their juices and the marinated salmon all at the same time. Stir gently for a few seconds. Warm the bowls, divide the noodles between them and ladle on the soup. Sprinkle the mint and spring onions over the top.

squid, pumpkin
& quail egg laksa

FOR SIX

1.8 litres (3 pints) fish stock
300g (11oz) pumpkin, cut into 6 wedges
200g (7oz) dried egg noodles
5 tablespoons sesame oil
12 baby squid, cleaned and washed
2 hot green chillies, finely sliced
6 lime leaves
6 baby aubergines, cut into quarters
12 fine slices of galangal
6 cloves of garlic, peeled and finely sliced
2 teaspoons Asian fish sauce
2 teaspoons tamari
2 tablespoons tamarind paste
50ml (1³/₄fl oz) fresh lime juice
6 quail eggs
Fresh coriander

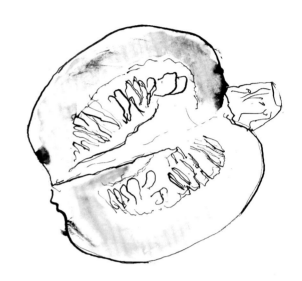

Bring the fish stock to the boil and add the wedges of pumpkin; boil the pumpkin until cooked and remove from the pot, but keep warm. Add the noodles to the pot and cook until done. Drain through a colander into a large bowl, keeping the liquid.

Return the emptied pot to the heat and heat 2 tablespoons of sesame oil until it begins to smoke. Add the baby squid and fry for 2 minutes, then remove and place with the pumpkin. Add the remaining sesame oil and, when hot, add the next five ingredients. Fry for 1 minute on a high heat, stirring well, then add the still warm fish stock, fish sauce, tamari and tamarind and turn to a simmer. After 5 minutes add the lime juice. While this is cooking, fry the quail eggs in a little cooking oil or a non-stick pan. Warm the six bowls and divide the noodles, pumpkin and squid between them, then ladle in the boiling liquid. Carefully place a quail's egg on top of each and add fresh coriander to taste.

This laksa not only looks fairly spectacular – with a tiny fried egg and squid tentacles floating on top – it also has a splendid array of clean tastes, all of which work well together. The sweetness of the pumpkin is a perfect foil to the chilli-fired heat. (For photograph see page 22.)

chicken,
buckwheat noodle,
roast garlic & coriander laksa

So what makes this so different from chicken noodle soup, you may well ask. Most chicken noodle soups I've had have been fairly insipid, but this relies on a gutsy, brown, roast chicken stock, full of dense buckwheat noodles and smoky roast garlic. Like all laksas, it's a filling meal in itself or, served in tiny bowls, it makes a wonderful starter with something simple like roast sea bass to follow.

FOR SIX MEAL-SIZED PORTIONS

3 boned chicken thighs
3 tablespoons sesame oil
18 cloves of garlic, not peeled
2 medium-sized onions, peeled and finely sliced
4 cloves of garlic, peeled and finely chopped
1 medium-sized carrot, peeled and finely diced
4 teaspoons peeled and finely grated fresh ginger
1 mild red chilli, finely sliced
6 tablespoons tamarind paste
2 litres (3^1/$_2$ pints) brown roast chicken stock (see page 164)
40ml (1^1/$_2$ fl oz) tamari
300g (11oz) buckwheat noodles, cooked (see page 109)
2 cups fresh coriander leaves

Preheat the oven to 200°C/400°F/Gas 6. Mix the sesame oil, chicken thighs and cloves of garlic together, spread them into one layer in an ovenproof dish and cook for 25 minutes.

When they're cooked, remove from the oven and turn it off. Drain the oil and juices into a large saucepan and put the chicken and garlic to the side. Heat the saucepan and add the onions, chopped garlic, carrot, ginger and chilli. Sauté over a moderate heat, stirring from time to time, until the onions have wilted. Add the roasted garlic, tamarind and chicken stock and bring to the boil, then simmer for 10 minutes. Cut the chicken thighs into small chunks, add them to the pan and warm through for a few minutes. Add the tamari and check for seasoning; add more if needed. Heat the buckwheat noodles under warm water and place in the preheated bowls. Ladle the laksa broth on to the noodles and sprinkle generously with the coriander leaves.

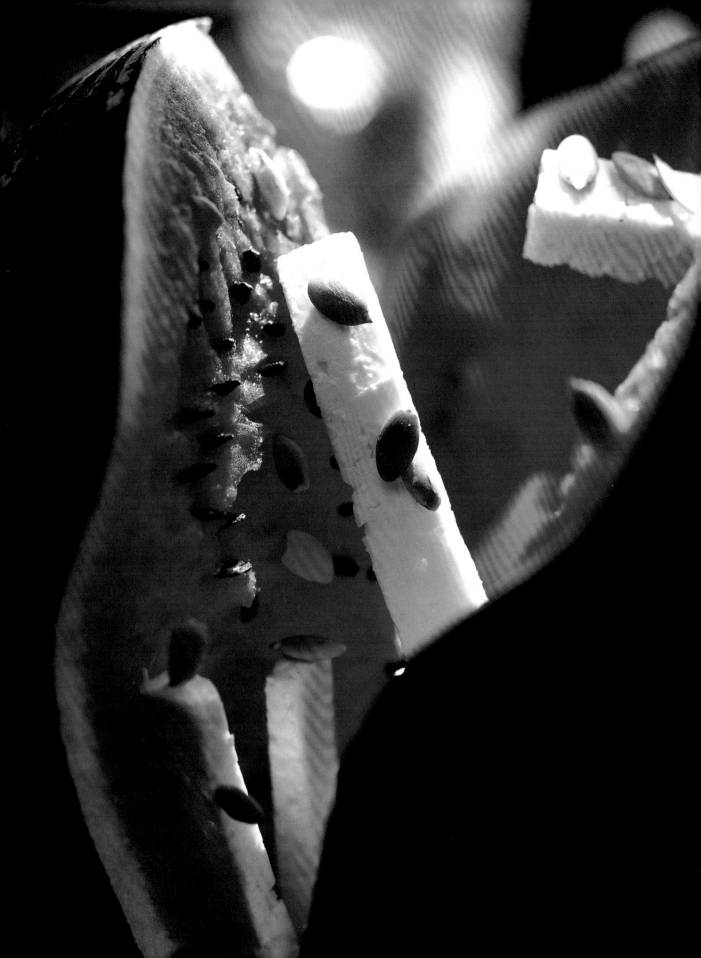

salads

vegetable salad with
sweet lemon & ginger
dressing

FOR FOUR AS A STARTER

6 large juicy lemons
1 clove of garlic
2cm (3/4 in) piece of peeled ginger
2 teaspoons honey
1 teaspoon caster sugar (unrefined is preferable)
Salt and white pepper
1 cup extra virgin olive oil
1 red pepper
6cm (2 1/2 in) piece of cucumber
1 cup bean sprouts
1/4 small red or savoy cabbage
1 small raw beetroot

Carefully peel three of the lemons, removing any pith. Cut the garlic
and ginger into eight pieces each and place with the lemon rind, honey
and sugar in a small saucepan. Juice the lemons (you need 1 cup of
juice) and add to the pan. Bring to the boil and reduce by half, add half
a teaspoon of salt and half a teaspoon of freshly ground white pepper,
allow to cool, then strain into a bowl and whisk in the olive oil.

Cut the pepper in half, remove the seeds and slice finely. Peel the
cucumber, slice in half lengthways and remove the seeds, then slice this
finely too. Rinse the bean sprouts in cold water and drain. Shred the
cabbage finely, removing the core first. Peel and finely julienne the
beetroot (you may want to wear gloves for this). Mix all the vegetables
together and toss with the dressing; leave for 20 minutes before
serving.

> The dressing for this salad is very versatile – it can also be
> used with grilled fish, steamed vegetables or cold poultry
> salads – and will keep in the fridge for five days.

salad of
watermelon,
feta
& toasted pumpkin seeds

As far as I know this salad has its roots in Israel. We serve it a lot at the Sugar Club, and though people are at first intrigued by the combination – finding it a little weird – they soon learn to love it. It's important to use an extremely ripe watermelon and good, salty feta. Quantities will depend on whether you want to serve it as a side salad or as a starter. (For photograph see page 30.)

FOR SIX AS A STARTER

1 cup pumpkin seeds
60ml (2fl oz) extra virgin olive oil
1.5kg (3¹/₄lb) watermelon
300g (11oz) sheep's feta
Crushed black pepper
3 juicy lemons

Preheat the oven to 180°C/350°F/Gas 4. Mix the pumpkin seeds with 20ml (³/₄fl oz) olive oil, spread on a baking sheet and toast in the oven for 8–12 minutes until golden. Don't burn. Cut the watermelon into chunks and peel it. If you have time remove the seeds, but it's not essential. Slice or crumble the feta over the watermelon, sprinkle with the toasted seeds and crushed pepper, drizzle with olive oil and serve with lemon wedges. Delicious!

grilled artichoke,
bean & caper salad

FOR FOUR STARTER PORTIONS

8 artichokes, stems removed at the base
Salt
400g (14oz) fresh beans, topped and tailed
2 lemons, cut into eighths
2 tablespoons coriander seeds

DRESSING

1 cup extra virgin olive oil
1/4 cup lemon juice
3 tablespoons seed mustard
2 tablespoons chopped fresh tarragon leaves
1/4 cup finely chopped fresh parsley
4 teaspoons capers, drained

Bring a large pot (large enough to hold the artichokes comfortably) of salted water to the boil. Add the beans, bring back to the boil and cook until *al dente* (about 2 minutes). Remove using a slotted spoon and refresh in cold water. Put the lemon pieces and coriander seeds into the pot and boil for 1 minute, then add the artichokes and return to the boil; cook on a rolling boil for 15–25 minutes – the time will depend on the size of the artichokes. To test, insert a skewer or sharp knife into the base and push into the centre: they are cooked when this can be done with only slight resistance. Remove the artichokes from the water, drain with the stem side up and leave to cool.

When the artichokes are cool enough to handle start pulling off the leaves. (The leaves are edible: you just hold the outer edge and scrape the fleshy bits off with your teeth, so they could be served as a separate dish.) Once all the leaves are removed pull out the centre 'choke', which is fibrous and inedible. Trim off any bits that look too chewy with a sharp knife. Lightly brush the hearts with some of the olive oil and grill until they start to colour a golden brown. This can be done on a skillet, over a barbecue or under a domestic oven grill. Mix all the ingredients for the dressing together well and pour over the artichokes, mix in the beans and leave to sit for at least 30 minutes before serving. Check for seasoning.

To my mind few things are more delicious – and fun – to eat than a boiled artichoke served at room temperature with a herb and mustard dressing. For some, however, this vegetable is as much of a challenge as a whole spiny lobster, with all the work of attacking the surrounding covering to get at the fleshy prize. So, to help your guests appreciate its taste, you can serve artichokes in a more prepared state, as here. The lemon and coriander in this recipe give an additional flavour and are a tip from fellow chef Jon O'Carrol. Artichokes prepared in this way also make an excellent topping for bruschetta.

new potato & pea salad
with mint dressing

This dish has a fresh taste that is just brilliant. It can be a starter on its own, but it's also good served with poached fish or chicken or cold roast leg of lamb. If you have time shell the peas yourself – your guests will be flattered! The dressing is enough for four main courses. This may seem excessive, but it's very tasty mopped up with bread.

FOR SIX STARTERS

1.5kg (3¹/₄lb) new potatoes, scrubbed
Salt
2 cups freshly shelled peas
1 large leek, finely sliced into rings and washed well

DRESSING

1 cup fresh mint leaves
³/₄ cup good vegetable salad oil
¹/₄ cup extra virgin olive oil
¹/₃ cup freshly squeezed lemon juice
1 tablespoon English mustard
2 teaspoons caster sugar
¹/₂ teaspoon sea salt
¹/₂ teaspoon cracked black pepper

Put the potatoes into a deep pot and cover well with cold water, add 1 tablespoon salt and boil. When they're almost cooked add the peas and leek, bring back to the boil and cook for a further 4 minutes. Test a pea – if it's ready, tip the contents of the pan into a colander and gently refresh under cold water for a few minutes.

There are two ways to make the dressing. The first and easiest is to place the oils and mint into a blender or small food processor and blend for 30 seconds, then add all the remaining ingredients and blend for another 10 seconds. The second is to shred the mint finely and mix it with the oils. Combine the lemon juice and mustard, then stir everything together.

While the vegetables are still warm toss with the dressing and leave to cool. The salad can be made the day before, but you may find that the peas lose their vivid colour. To avoid this, cook them separately and mix in an hour or so before serving the dish.

broad bean &
pancetta salad

Few things are more rewarding than to sit in the sun with a large bowl of pods and watch the pile of beautiful, shiny beans get larger and larger as you shell them. The silky inner texture of the pods is a delight. For this recipe you only need to extract the beans from the pod – you don't need to skin the beans themselves. Pancetta is rolled, cured pork belly and is available from good Italian delicatessens. Ask them to slice it thinly for you. If you are unable to find pancetta, use diced smoked streaky bacon instead.

FOR FOUR

4 cups podded broad beans
50ml (1³/4fl oz) olive oil
200g (7oz) sliced pancetta, coarsely shredded
1 medium-sized white-fleshed onion, peeled and finely sliced
2 cloves of garlic, peeled and sliced
1 cup fresh flat-leaf parsley leaves
50ml (1³/4fl oz) balsamic vinegar
Black pepper

Heat the oil in a deep frying pan and add the pancetta. Cook on a high heat until it begins to crisp, then add the beans and sauté until the skins start to burst, stirring gently every few seconds. Add the onion and garlic and cook for another minute, stirring well. Remove from the heat and stir in the parsley. Just before serving mix the vinegar through the salad and grind some black pepper over it.

roast red onion, chicory, croûton & asparagus salad

FOR SIX STARTER-SIZE SALADS

3 medium-sized red onions, peeled and sliced into 5mm ($^1/_5$in) rings

150ml (5 fl oz) extra virgin olive oil

100ml (3$^1/_2$ fl oz) balsamic vinegar

$^1/_2$ teaspoon sea salt

$^1/_2$ teaspoon coarsely ground black pepper

6 slices stale bread, crust removed and cut into rectangles

2 tablespoons olive oil

4 heads of chicory

24 asparagus stems

Turn the oven to 180°C/350°F/Gas 4. Put the sliced onions, extra virgin olive oil, balsamic vinegar, salt and pepper into a ceramic ovenproof dish and seal tightly with foil. Use a dish that's just large enough to hold the onions. Bake in the oven for 1 hour, remove the foil and cook for another 15 minutes, then remove and cool completely. Drop the temperature to 150°C/300°F/Gas 2 and lay the bread slices on a baking sheet. Brush them with the olive oil and bake until golden on both sides, then remove from the oven to cool.

Cut 2cm (¾in) from the base of each chicory head and separate the leaves, wash and drain. Bring a large pot of salted water to the boil. Holding both ends of each asparagus stem, bend them towards each other until the stem snaps. Discard the lower end (or use in a vegetable stock). Drop the asparagus tips in the boiling water, return to the boil and cook for 1 minute. Drain and refresh in cold water. Mix the onions and their cooking juices, the chicory and asparagus together. Serve on a plate with the croûtons on top.

This is a perfect summer salad full of simple flavours and textures. The roast red onions are useful to have around and keep for up to a week in the fridge. Depending on what you're using them for you can add an assortment of extra flavours, such as chillies, herbs or garlic. Use any bread for the croûtons and cut them into the shape you want; here I've suggested wholemeal.

salad of
pickled gooseberries,
red cabbage, hazelnuts & coriander

FOR TEN STARTERS

1/2 red cabbage

500g (18oz) gooseberries, washed well

500ml (18fl oz) cider vinegar

1 litre (1 3/4 pints) water

500g (18oz) light brown soft sugar

4 star anise

2 teaspoons coriander seeds, lightly crushed

1/2 cup finely chopped peeled ginger

2 garlic cloves, peeled and finely chopped

1 cup shelled hazelnuts, roasted and peeled

2 bunches fresh coriander, leaves picked off the stalks

150ml (5fl oz) extra virgin olive oil

Slice the red cabbage as fine as you can, removing the thick core first. Pack it tightly into a preserving jar and place in a sink or bowl as deep as the jar. Pack the gooseberries into another jar. Pour enough hot water into the sink or bowl to come at least one-third of the way up the jars – this will prime the glass and prevent it from shattering when you pour in the hot pickle liquid. Put the next seven ingredients into a pot and boil for 2 minutes; strain half into the cabbage and then, without straining, pour the rest on to the gooseberries. The liquid should cover the contents; if it doesn't, bring some more water and vinegar (at a ratio of 2 to 1) to the boil and top up the jars. Gently wobble the jars to release trapped air, then seal quickly and leave to go cold in the bowl or sink. Put in the fridge and leave for at least a week, turning the jars upside down every two days.

To serve, mix half the cabbage and half the gooseberries with the hazelnuts, coriander and olive oil, add 200ml (7fl oz) of the liquid from the gooseberries and mix well.

This is a dish that I first made at the original Sugar Club in Wellington, and it's still one of Ashley and Viv's favourites. Pickling was something my grandmother got me interested in as a kid. Years later, as I began to eat at Japanese restaurants, I realised the role that pickles play in their cuisine as an aid to digestion and a stomach teaser. The pickles here are best made at least one week in advance and will keep for several months in the fridge.

pickled vegetable,
seaweed & sesame salad

This is a salad that stands alone as a course in its own right, but it is also excellent topped with grilled fish – especially an oily fish like mackerel or salmon. In this recipe I use arame seaweed, but you can use any seaweed you fancy. One thing to aim for is a good contrast in colours, sweetness and acidity, so choose an assortment of vegetables and spices to suit your taste. You'll need to make the pickled vegetables a few days in advance to get the benefit of the pickling. Stored in a jar in the fridge they will keep for three months.

FOR SIX STARTERS

1 medium-sized carrot, peeled and finely julienned

1 10cm (4 in) piece of mooli (daikon), peeled and finely julienned

1/2 cucumber, seeded and finely sliced into rings

1 red pepper, seeds removed and finely julienned

1 cup bean sprouts

3 tablespoons finely grated peeled ginger

300ml (11fl oz) cider vinegar

600ml (1 pint) water

150g (5oz) caster sugar

1 teaspoon salt

1/2 teaspoon finely ground star anise

25g (1oz) dried arame seaweed

50ml (1 3/4 fl oz) sesame oil

4 tablespoons sesame seeds, toasted

1 cup finely sliced spring onions

Mix all the vegetables and the ginger together. Bring the vinegar, water, sugar, salt and star anise to the boil. Fill a glass preserving jar with hot tap water and leave it to warm for a few minutes, then tip the water out and pack the vegetables in. Pour the boiling vinegar over the vegetables and seal the jar while hot. Leave until completely cold, then place in the fridge for a few days.

Soak the arame seaweed in warm water for 3 hours and then drain, keeping half a cup of the liquid. Heat the sesame oil in a pan and fry the arame for a few minutes over moderate heat, add the soaking liquid and cook until it has evaporated, then leave to cool. To serve, mix half the pickled vegetables and arame with the sesame seeds before adding 100ml (3 1/2 fl oz) of the pickling liquid and the spring onions.

warm salad of
chickpeas,
chilli, feta & garlic

FOR A LARGE BOWL OF SALAD OR EIGHT LARGE STARTER-SIZE PORTIONS

300g (11oz) dried chickpeas (or use two 450g/1lb tins of precooked)
Kombu or bicarbonate of soda (see note at bottom of page)
200ml (7fl oz) olive oil
3 red chillies, de-seeded and finely sliced
12 cloves of garlic, peeled and roughly chopped   *1 cup roughly crumbled feta*
3 red onions, peeled and finely sliced   *1 cup sliced spring onions*
150ml (5fl oz) cider vinegar   *1 cup fresh mint leaves*
1 cup roughly chopped fresh coriander leaves   *1/2 cup extra virgin olive oil*
1 cup fresh flat-leaf parsley leaves   *Salt and pepper*

Soak the dried chickpeas overnight in plenty of cold water. The next day drain and rinse them, then decant into a deep pot. Add enough cold water to submerge them by at least 5cm (2in) and bring to the boil before turning down to a rapid simmer. Skim off any scum that forms during cooking. If adding kombu or bicarbonate of soda, add it once the chickpeas come to the boil. Keep topping up with boiling water so the peas don't dry out; test them by eating one – it should be firm and a little grainy but with no crunch. Depending on the size of chickpea you will need to cook them for 1–2 hours. When they're done, drain well and rinse under hot water.

Heat the olive oil in a saucepan, then add the chilli, garlic and onions and cook over a high heat for 5 minutes, stirring quite frequently to prevent sticking. Add the vinegar and boil until it has evaporated (about 2 minutes). Put the chickpea and onion mix in a large bowl while still warm, add the remaining ingredients and mix really well. Leave to sit for 15 minutes before giving it another stir and test for seasoning.

As with butter beans, two rules apply when cooking dried chickpeas: add salt at the end and skim off the scum. Chickpeas take a long time to cook, but the process can be speeded up by adding half a sheet of kombu seaweed or a teaspoon of bicarbonate of soda per 500g (18oz) chickpeas while boiling, though the latter may cause a loss of nutrients. This is a delicious salad in its own right, but served with grilled squid or roast lamb it's even better. Using precooked chickpeas from a can will save time – just remember to rinse them well before using.

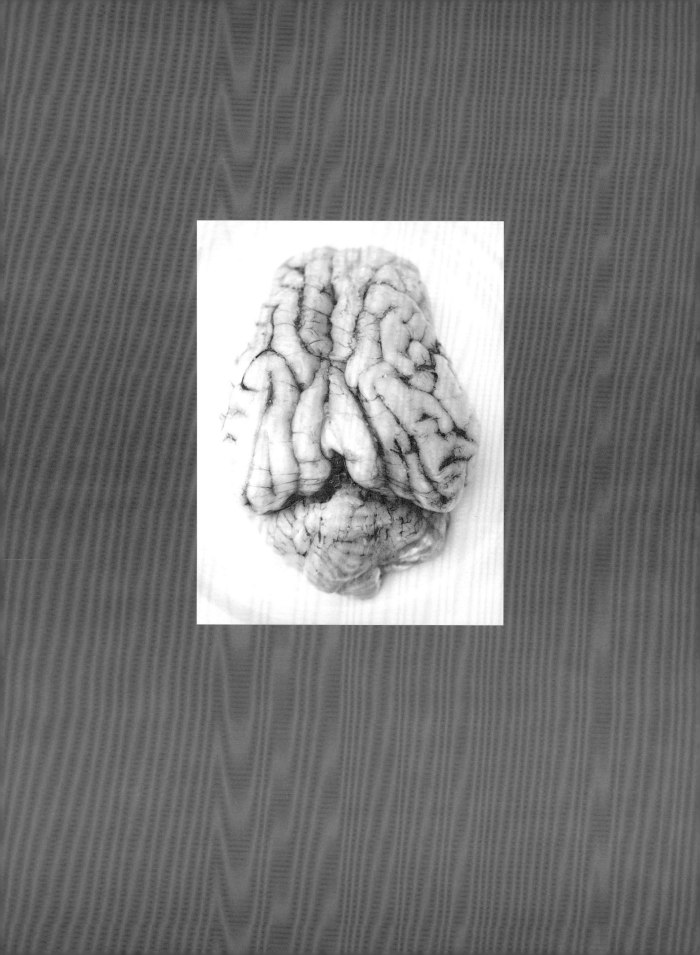

offal

grilled ox tongue
with pesto & tomato

Gran often had a jellied tongue in her larder and it was one of the few things she made that I didn't like. It was years later, when I made my first brawn, that I discovered the joys of this meat. Here the smoky richness of the grilled tongue is complemented by the acidity of the tomato and the flavour-packed pesto. Make the pesto as on page 165, adding 3 tablespoons of fresh tarragon to the recipe. One ox tongue will give you enough for 12 generous starters or six main courses.

1 ox tongue (buy pickled and rinse under cold water for a few minutes)
1 carrot, peeled
1 medium-sized onion, cut into quarters with the skin on
1 head of garlic, cut in half
6 bay leaves
12 sage leaves
1 8cm (3in) rosemary branch
2 teaspoons black peppercorns
100ml (3 1/2 fl oz) vinegar
1–2 very ripe tomatoes per person
Salt and pepper
Pesto (see page 165)

First, cook your tongue. Put all the ingredients except the tomatoes, seasoning and pesto into a large pot and cover with cold water. It's important that the tongue floats, so make sure there is enough water. Bring to the boil and skim off any foam that forms on the surface, then turn down to a rapid simmer and cover. Cook for at least 90 minutes and check from time to time that the water hasn't reduced too much; top up with hot water if it has. To test if the tongue is cooked try to peel off some of the skin: it should come off easily. Once it is done, remove the pot from the heat and leave to cool. When cooled, take the tongue and peel away the tough skin with your fingers. Trim off any bits of fat and sinew and cut into 1cm (1/3in) slices.

Heat up a grill. Cut the tomatoes into 1cm (1/3in) slices and season lightly. Grill the tongue until it starts to colour, then turn over and repeat on the other side. Serve the warm tongue on the tomato slices with a generous dollop of pesto to top it off. You may also want to serve a bowl of hot mustard on the side.

pan-fried lambs' brains
on toast with pear & date chutney

I love lambs' brains, and it's the combinations of this dish – textural as well as tastewise – that make it a firm favourite with offal lovers, myself included. With any brain dish you make proceed as in this recipe up to the poached stage. In some restaurants hours are spent peeling the outer membrane from the brain, which I think is a waste of time. For the toast use any good, dense bread that's slightly stale. You'll find chutney recipes in the section on relishes, but any good chutney works just as well.

FOR SIX LARGE STARTERS

9 lambs' brains

1 red onion, cut into quarters with the skin on

1 carrot, peeled and cut in half lengthways

1 stalk celery, cut in half

4 bay leaves

4 cloves of garlic, flattened with the skin on

100ml (3½fl oz) cider vinegar

Salt and cracked black pepper

150g (5oz) unsalted butter

30ml (1fl oz) balsamic vinegar

200ml (7fl oz) brown roast chicken stock (see page 164)

6 slices from a large loaf of bread

Chutney

Make sure the brains you buy have been removed from the skull in one piece. Fill a bowl with warm water and wash them gently, removing any small pieces of adhering bone. Take the brains out, fill the bowl with fresh water and leave the brains to sit in it for 30 minutes.

Put the onion, carrot, celery, bay leaves, garlic, cider vinegar, 2 teaspoons salt, 1 teaspoon pepper and 300ml (11fl oz) cold water into a large, deep pot and boil for 2 minutes, then allow to cool completely. Add the brains to the pot and fill with enough cold water to cover them by 3cm (1in). Put the pot back on the stove and bring to the boil over a high heat, stirring gently from time to time to stop the brains sticking to the bottom. Once they come to the boil, turn down the heat and simmer for 5 minutes. Then transfer the pot to the sink, turn the cold tap on slowly and run the water gently over the brains until it runs off cold. Take one brain out at a time and separate it into its two lobes, discarding the small round-shaped bit in the middle; also discard the vegetables. Put the divided brains into a bowl and rinse them gently in cold water, then drain.

Heat up a large frying pan and add the butter; keep on a high heat and fry until the butter browns, then add the brains. Fry them for about 4 minutes until golden brown, then turn over and fry for another 4 minutes. To make the sauce, add the balsamic vinegar, stock and ½ teaspoon each of salt and pepper and simmer for 2 minutes. Toast the bread and spread a thick layer of chutney on top. Put three lobes of brain on each piece of toast and spoon the sauce over them.

grilled kidneys
with tamarind & garlic

Kidneys were the last kind of offal I developed a liking for – I'd always associated their taste with brussels sprouts. In fact I still think of brussels sprouts as the kidneys of the vegetable world. In this recipe I use lamb kidneys, but the same recipe could be applied to veal kidneys. Buy your kidneys with the suet (the fatty covering) removed. The dish is best served simply, as a starter on toast or as a main course on top of mustard mash and steamed spinach or kale.

FOR SIX STARTERS

12 lambs' kidneys
50ml (1³/4fl oz) sesame oil
Salt and pepper
6 cloves of garlic, peeled and finely chopped
1 small red onion, peeled and finely sliced
¹/2 teaspoon finely ground coriander seeds
150ml (5fl oz) tamarind paste
250ml (9fl oz) brown roast chicken stock (see page 164)

Peel the outer membrane from the kidneys, cut them in half lengthways and remove the fatty core. Toss in a little of the sesame oil and season lightly. Turn the grill on high. Heat the remaining sesame oil in a saucepan and, once it's smoking, add the garlic and stir well, then add the onion and coriander seeds. Sauté until the onions have wilted, then add the tamarind and stock. Bring to the boil, test for seasoning, and leave to simmer while you grill the kidneys.

Put the kidneys on a rack with the cut side up and grill for 3 minutes with the rack 4–5cm (1½–2in) from the grill. Turn and cook for a further 3 minutes. Cook them to the point you think best – people's tastes in kidneys vary from rare to rubbery to well done. Serve on a warm plate with the hot sauce on top.

sautéed
duck livers
with fenugreek & chilli
on potato & coriander salad

I recently found a jar of fenugreek seeds in my cupboard – a leftover from growing fenugreek sprouts in the summer. It's the spice that gives curry powder much of its aroma. It can be quite bitter if too much is used, so go carefully. This dish is very simple and, like most good food, it relies on really fresh ingredients. You can replace the fenugreek with ground star anise or cumin, and chicken livers are almost as good as duck livers. The former need less cooking, but, like duck livers, they're best eaten a little pink in the middle.

FOR SIX STARTERS

12 new potatoes, boiled, skins rubbed off, then sliced in half
1 small red onion, peeled and finely sliced
1 cup fresh coriander leaves
50ml (1 3/4 fl oz) extra virgin olive oil
2 cloves of garlic, peeled
1 green chilli
1 teaspoon fenugreek powder
50ml (1 3/4 fl oz) sesame oil
12 duck livers, trimmed of any sinews or fibres
300ml (11 fl oz) chicken stock
1 teaspoon balsamic vinegar
1 teaspoon Asian fish sauce

Mix the potatoes, red onion, coriander and olive oil and divide between six plates. In a pestle and mortar grind the garlic, chilli and fenugreek to a fine paste along with 20ml (3/4 fl oz) of the sesame oil. Heat a frying pan until it's very hot and add the remaining sesame oil. When the oil is smoking add the livers and cook for 1 minute, then add the paste and turn the livers over. Fry for another minute, then add the stock, balsamic vinegar and fish sauce. Bring to the boil and after 1 minute remove the livers and place them on the potato salad. Keep the sauce boiling until it starts to thicken, then pour it over the livers.

sautéed
lambs' sweetbreads,
lemon & fennel seeds on grilled polenta

The first time I ate sweetbreads they were cooked in cream and veal stock and their appeal totally evaded me. It all seemed too creamy and soft. I like to cook them with something a little sharp in taste or to deep-fry them, either in batter or crumbed, to give them a good texture. You can also use veal sweetbreads, though these are horribly expensive. If you do, prepare them in the same way but poach them for longer, and when you come to sauté them, slice them on an angle into pieces.

FOR SIX STARTERS

1 carrot, peeled and cut into quarters

1 onion, peeled and sliced

4 cloves of garlic, roughly chopped with the skin still on

2 bay leaves

1 lemon, cut into 6 slices

Small bunch fresh thyme

1 teaspoon fennel seeds

1 litre (1 3/4 pints) water

50ml (1 3/4 fl oz) vinegar

1kg (2 1/4 lb) lambs' sweetbreads

50ml (1 3/4 fl oz) olive oil

50g (1 3/4 oz) unsalted butter

1 red onion, peeled and finely sliced

2 teaspoons fennel seeds

2 lemons, grate the zest, then juice them

100ml (3 1/2 fl oz) brown chicken stock

1/2 cup chopped fresh parsley

POLENTA

125g (4 1/2 oz) polenta grains,
    sieved to break up any lumps

1 red onion, peeled and finely diced

1 teaspoon finely chopped rosemary

2 cloves of garlic, peeled and finely chopped

30ml (1 fl oz) olive oil

1 teaspoon salt

500ml (18 fl oz) chicken stock

Put all the ingredients from the carrot to the vinegar in a deep pot and bring to the boil, boil for 3 minutes, then gently add the sweetbreads and stir well. The sweetbreads must be well covered with liquid; if not, top up with hot water. Bring back to the boil and simmer for 5 minutes before removing the pot and leaving the sweetbreads to cool in the poaching liquor. Once cooled, drain them and pull any fatty bits off, then lay them on a plate and sit an identical plate on top. Place some weights (e.g. cans of food) on the upper plate and leave for 2 hours.

Meanwhile, make the polenta. Sauté the onion, rosemary and garlic in the olive oil until the onion softens. Add the salt and stock and bring to the boil. Turn to a simmer, then slowly trickle the polenta in, whisking as you go to prevent lumps forming. When all the polenta has been added, remove the whisk and stir with a spoon. Return to the boil and simmer for 1 minute. Pour the polenta on to an oiled plate and leave it to cool.

When you're ready to serve the dish, cut the polenta into wedges and place on an oiled tray in a hot oven (220°C/425°F/Gas 7) for at least 5 minutes. Heat the olive oil and butter in a frying pan over a high heat until the butter starts to brown. Add the sliced onion and fennel seeds and sauté, stirring, for 1 minute. Add the sweetbreads and fry until golden on both sides, then add the lemon juice, zest and chicken stock. Bring to the boil, taste for seasoning and stir in the parsley. Spoon the sweetbreads on to the hot wedges of polenta and enjoy.

baked lambs' brains
with feta, smoked paprika & labne

Here is a delicious combination of tastes – salty, spicy, sharp and rich – that makes a great starter for the winter. It was a favourite of Frances Lang, a photographer friend. As she said, 'It's so you, Peter – such an odd combination, yet it works so well.' Naturally, this dish is dedicated to her. You'll find the labne on page 165. Prepare the brains as in the recipe on page 48.

FOR SIX STARTERS

300ml (11fl oz) labne (see page 165)
6 poached lambs' brains, divided into 12 lobes; cut each lobe in half
300g (11oz) feta, coarsely grated
Freshly ground black pepper
2 teaspoons sweet smoked paprika
50ml (1³/4fl oz) extra virgin olive oil

Turn the oven to 200°C/400°F/Gas 6. Lightly oil six shallow oven-proof ramekins, put 2 teaspoons of labne in the bottom of each and place four halves of brain on the labne. Sprinkle the feta over them and grind some black pepper on top of this. Mix the remaining labne with the paprika and olive oil and spread lightly over the brains. Place the ramekins on a baking tray and bake for 6–10 minutes until the labne begins to bubble and the brains are hot. Serve at once – but be warned, they will be scalding.

shell-fish

grilled crayfish
with coriander butter & tomato salsa

FOR TWO MAIN COURSES

1 1kg (2¼ lb) crayfish, killed and halved (see below)
150g (5oz) salted butter at room temperature
½ cup fresh coriander leaves
1 teaspoon finely ground coriander seeds
½ teaspoon finely grated lime zest
½ teaspoon cracked black pepper

TOMATO SALSA

3 tomatoes, finely diced
½ cup finely sliced spring onions
30ml (1fl oz) lime juice

Make the salsa before cooking the crayfish. Just mix the tomato dice with the spring onions and lime juice and leave for 10 minutes.

Heat up an overhead grill until it's really hot. Put the butter, coriander leaves and seeds, lime zest and pepper in a food processor and purée for 30 seconds. Spread half the butter mix over the two halves of the crayfish and grill for 3 minutes. Then spread the rest of the butter over the crayfish and grill for a further 3 minutes. Be careful that the butter doesn't flare up under the grill, as it might, and make sure the crayfish is a little rare – it will continue to cook a little as it sits. Spoon the salsa over the cooked flesh and eat straightaway. Use the leftover shells to make a bisque.

In New Zealand we went crayfishing a lot as a family. We caught them at Cape Palliser in cray-pots or when snorkeling up at the Coromandel at Christmas. In Europe the term for this marine shellfish is 'spiny lobster'. For this recipe you can use either crayfish or lobster. It is essential that they are alive when you buy them, as this way you'll know they're fresh. To kill them humanely weigh them down in a large bucket of very cold tap water for 4 hours and they will drown (they cannot survive in non-salt water). Hold them flat on a chopping board with their tails out straight and, starting at the head, cut them in half lengthways with a large, sharp knife. Be really careful as you do this because a slip could result in serious damage to your fingers.

wok-fried scallops
with pak choy & peanut sauce

Wok-frying imparts a slightly smoky taste to a lot of food – it's the high heat that does it. Scallops especially improve with this method (or with char-grilling), but you can also smoke them as in the venison recipe on page 111, taking care to not overcook them. Scallops are always best eaten rare in the middle. If you can't find pak choy use any Chinese green or finely shredded savoy cabbage.

FOR SIX STARTERS

18 large scallops, shelled

Sesame oil

600g (1¹/₄ lb) pak choy, washed and chopped into 6cm (2¹/₂ in) lengths

10ml (¹/₃ fl oz) Asian fish sauce, diluted with 30ml (1 fl oz) water

¹/₂ cup fresh coriander leaves

3 fresh juicy limes

PEANUT SAUCE

³/₄ cup toasted peanuts

2 cloves of garlic, peeled

2 teaspoons finely grated ginger

¹/₂ green chilli, roughly chopped

2 tablespoons brown sugar

300ml (11 fl oz) unsweetened coconut milk

Trim any muscles or membranes from the scallops, wash gently in cold water and dry well. Mix with just enough sesame oil to coat them and leave to sit at room temperature for 15 minutes.

To make the sauce, put the peanuts, garlic, ginger, chilli and sugar in a food processor and purée to a coarse paste, then briefly mix in the coconut milk.

Heat up a wok until it's really hot. Add nine scallops and cook for 1 minute on each side, tossing gently for a few seconds at the end. Transfer them to a warm bowl and put the wok back on the heat. Do the next nine scallops the same way. When finished, put a few tablespoons of sesame oil in the wok and, once it's really hot, add the pak choy and toss every 10 seconds for half a minute. Add the diluted fish sauce and fry for a further 20 seconds. Now add the peanut sauce and bring to the boil. Add the scallops and just warm them through, stirring once. Serve with the coriander on top and some lime segments.

black bean, sweetcorn, cucumber & scallop broth

FOR SIX MAIN COURSES

30 shelled scallops with roe (or coral), trimmed of muscle or
*   membrane and gently rinsed under cold water*
Sesame oil
2 leeks, finely sliced and washed to remove grit
2 cloves of garlic, peeled and finely chopped
2 teaspoons finely grated ginger
1/4 cup salted black beans (found in most Asian food stores)
800ml (28fl oz) fish stock
30 baby sweetcorn
1 15cm (6in) piece of cucumber, seeds removed and cut into fine juliennes
1/4 cup lime juice
1/2 cup finely sliced spring onions

Heat up a deep pot and add a few tablespoons of sesame oil. When it's smoking put in the leeks, garlic and ginger and sauté for 3–4 minutes, stirring occasionally. Add the black beans and fish stock and bring to the boil, then turn to a simmer.

Heat up a heavy frying pan and, when it's hot, add a few tablespoons of oil. Fry the sweetcorn until golden brown all over (don't overcook or it will burn and its texture will soften) and add to the broth. Wipe out the frying pan with kitchen paper, heat it up again and put in a little more oil. Gently add the scallops and fry on a high heat until they're golden on each side. Put these in the broth with the cucumber and turn the heat off. Add the lime juice and mix in gently. Taste for seasoning before serving in flat soup bowls sprinkled with the spring onions.

The key thing about this dish is its combination of tastes and textures. The saltiness of the black beans, Asian in origin, contrasts with the sweetness of the corn and scallops and the bitterness of the scallop roe. Many chefs recommend discarding the roe, but ignore their advice as it's delectable. The crunch of the corn and cucumber contrasts with the softness of the scallops. Use this recipe as a basis to start from and add whatever you think will go well. If you don't have any fish stock to hand, use a light chicken or vegetable stock.

When prawns are described as 'green' it means simply that they're raw. Green prawns can be bought in all sorts of ways – with their heads or shells on or off and frozen or fresh. The main thing is not to buy precooked as these tend to be quite inferior. For the recipe opposite you can use prawns out of their shells, but it is the crisp shell that is so delicious to munch on. Seriously! When really fresh there's nothing tastier than a whole prawn, eaten head and all. This was something I discovered throughout Asia whenever I ate barbecued prawns. The recipe is a favourite of my sister Tracey – we had it for Christmas lunch in 1996 in the rainforest of northern New South Wales.

pan-fried
green prawns
with ginger & garlic

FOR SIX STARTERS

18 medium-sized green prawns, shells and heads intact
8 cloves of garlic, peeled
3 thumbs of fresh ginger, peeled
1 cup sesame oil
1 teaspoon salt
1/2 cup fresh coriander leaves
1/2 cup lime juice

Rinse the prawns with cold water and dry them well in a cloth. Put the remaining ingredients, except the lime juice, into a food processor and purée to a paste. Toss the prawns in this, cover with clingfilm and keep in the fridge for at least 6 hours.

When you're ready to cook the prawns, heat a heavy frying pan (preferably non-stick) until it is extremely hot – the heat is needed to make the shells go crisp. Carefully add as many prawns as will fit comfortably at one time, shaking off excess marinade as you go. Fry them for 2 minutes on each side until they are slightly blackened and crisp. When they're all done add the remaining marinade to the pan and fry for a minute. Then add the lime juice and, when it comes to the boil, spoon the marinade over the cooked prawns.

Eat with your fingers and I do urge you to eat the whole thing – you'll be deliciously rewarded! The prawns can be served with some crusty bread or boiled rice to mop up the juices.

crispy deep-fried
prawns & okra
with mango salsa

FOR SIX GENEROUS STARTERS

18 raw prawns
12 finger-sized okra, unblemished
Vegetable oil for deep-frying

BEER BATTER

1 1/2 cups plain flour
1 teaspoon salt
1 teaspoon dried chilli flakes
1 teaspoon brown sugar
1 teaspoon baking powder
450ml (3/4 pint) beer

MANGO SALSA

1 large ripe mango
1/2 cup finely sliced spring onions
1/2 red chilli, seeds removed and finely diced
1/4 cup fresh lime juice
1 tablespoon sesame oil
1/2 cup finely sliced fresh basil leaves
Salt and pepper

Mix the dry ingredients for the batter together and then whisk in the beer, making sure there are no lumps. Let the batter rest for 20 minutes.

To make the salsa, peel the mango and cut the flesh from the stone in two halves. Dice the flesh and mix it with the remaining ingredients. Rest for at least 10 minutes before using.

Take the shells off the prawns but leave the tail end with its little flipper intact. This will make them easier to hold when you put them in the fryer and it's also deliciously crunchy to eat. Remove the intestinal tract by running a small knife along the top of the prawns and pulling out anything that isn't white flesh. Cut the okra in half lengthways, put with the prawns into the batter and toss well. In a deep-fryer, or deep pot, heat up some vegetable cooking oil until it is just smoking (180°C/350°F). Cook the okra by lowering them individually into the hot oil; once they've turned a golden brown remove from the oil and drain on absorbent paper. Keep in a warm place while you cook the prawns in the same way. Serve with the salsa on the side.

This is a great mix of all the tastes and textures I love: sweet and tangy, savoury and salty, soft and crunchy. Deep-fried foods seem to appeal to us all and feature in cuisines the world over. The first time I saw okra growing was in a cousin's garden in Conway, Arkansas, when my partner Michael and I were visiting *en route* to England in 1989. It is a member of the cotton family and one of the most beautiful plants I've seen. That night – as we feasted on okra fritters and chicken bake – we were set upon by an amazing tornado that wreaked havoc around the state. This must be why I associate okra with exciting things like chilli, spice – and storms!

steamed mussels
with
tomato, basil, cumin & olive oil

Mussels are a fun dish to eat and they're extremely tasty as well. The flesh is meaty and will hold its own with many flavours. The following simple recipe is a particular favourite of mine, but it does rely on using the ripest, best tomatoes. Make sure too that the mussels are fresh, and clean them thoroughly. Pull off any hairs and scrape away barnacles with the back of a blunt knife. Then wash them well in cold water, discarding any that are already open or have cracked shells. To store, put the washed and drained mussels in a bowl, cover with a wet cloth and place in the fridge. Some important rules: don't keep them long as they pass their sell-by date quickly, don't store them in water and never cover them with clingfilm as they will suffocate – like oysters, mussels are sold alive.

FOR SIX STARTERS

2kg (4^1/$_2$ lb) cleaned mussels
600ml (1 pint) dry white wine
1kg (2^1/$_4$ lb) ripe tomatoes, cut into 1cm (^{1}/$_3$ in) dice
6 teaspoons cumin seeds, toasted
1 cup fresh basil leaves
1 cup fresh flat-parsley leaves
1 cup sliced spring onions
^{1}/$_2$ cup olive oil
2 teaspoons cracked black pepper

Bring the wine to the boil in a large pot with a tight-fitting lid. Mix the tomato with the cumin, basil, parsley, onions, olive oil and pepper. Add half of this to the boiling wine and tip all the mussels on top. Pour on the remaining tomato mixture and give it all a stir. Bring back to a full boil with the lid on and don't look for 5 minutes. Then tip the mussels into bowls, discarding any that refuse to open as they may not be good. Serve with finger bowls – it's the best way to eat them.

grilled scallops
with sweet chilli sauce
& crème fraîche

FOR FOUR PEOPLE

12 large diver-caught scallops, trimmed

Sesame oil

Salt and pepper

Watercress leaves

1/2 cup crème fraîche

SWEET CHILLI SAUCE

10 cloves of garlic, peeled

4 large red chillies, stems removed

3 thumbs of fresh ginger, peeled and roughly chopped

1 thumb of galangal, peeled and roughly chopped

8 lime leaves

3 lemon-grass stems; remove the two outside leaves,
*   discard the top third of the stem and finely slice the remainder*

1 cup fresh coriander leaves

1 1/2 cups caster sugar

100ml (3 1/2 fl oz) cider vinegar

50ml (1 3/4 fl oz) Asian fish sauce

50ml (1 3/4 fl oz) tamari

Put the first seven ingredients of the chilli sauce in a food processor and purée to a coarse paste. Put the sugar in a saucepan with 4 tablespoons of water and place on a moderate heat, stirring well until the sugar dissolves. When it has, remove the spoon and turn the heat up to full. Boil for 5–8 minutes and do not stir until it has turned a dark caramel colour (but don't allow it to burn). Now stir in the paste, bring the sauce back to the boil and add the last three ingredients. Return to the boil and simmer for 1 minute. Leave it to cool before eating.

Lightly oil the scallops with sesame oil and season, then grill each side on a char-grill, overhead grill or skillet for 90 seconds. Sit them on a bed of watercress, put a dollop of crème fraîche on top and drizzle generously with sweet chilli sauce.

This dish – one of my all-time favourites – is now synonymous with the London Sugar Club. I first put it on the menu in July 1995, and it has only come off when storms prevented divers collecting scallops. It's also the dish that most reviewers choose when they visit! The crème fraîche softens the strong flavours of the sauce, which in turn cut through the richness of the scallops. The chilli sauce recipe makes more than you need, so keep the surplus in the fridge for other dishes.

crab
& potato cakes
with salted cucumber

FOR SIX STARTERS

1 cucumber, julienned

¹/₂ cup coarse sea salt

3 tablespoons unrefined caster sugar

80ml (3 fl oz) lemon juice

400g (14oz) sweet potato, peeled and grated

800g (1³/₄lb) baking potato, peeled and grated

1 egg, beaten

50g (1³/₄oz) cornflour

1 teaspoon salt

1 teaspoon cracked black pepper

1 cup finely sliced spring onions

Cooking oil (peanut works well)

600g (1¹/₄lb) crab meat

¹/₂ cup fresh coriander leaves

120ml (4 fl oz) light olive oil

Mix the cucumber and coarse salt together in a bowl, tip into a colander and leave to drain into a bowl for 1 hour. Then rinse the cucumber well under lots of cold water and mix in a bowl with the sugar and half of the lemon juice. Leave to sit for 30 minutes.

Squeeze as much liquid from the grated potatoes as you can (discard the liquid) and mix with the egg, cornflour, salt, pepper and spring onion. Divide the mix into twelve and make into 'cakes' about 8mm (¼in) thick. Heat up a heavy non-stick frying pan to a moderate heat, brush lightly with a little cooking oil and fry each cake until golden on both sides (flip them over with a spatula, taking care not to break them). A convenient way to finish the potato cakes is to place them in a hot oven for a few minutes on a lightly oiled tray. Keep them warm after making or reheat just before serving.

Mix the crab meat and coriander. Mix the remaining lemon juice with the olive oil. Place a warm potato cake on each plate, put a sixth of the crab mixture on them and put another cake on top. Pile the cucumber mixture on top of that, drizzle with the lemon dressing and eat.

For this recipe either use pre-prepared crab meat or cook your own crabs and remove the flesh. The latter is time-consuming and requires a pair of shell-crackers, but it's worth it as the meat will be that much better. The combination of normal and sweet potatoes offsets the saltiness of the cucumber and works extremely well. It is important to rinse the cucumbers thoroughly or they may be too salty. Grate the potatoes just before you need them as they discolour if left for too long.

new zealand
green-shell
mussels
with mooli, hijiki, chilli & rocket

For this recipe I specify these native mussels from New Zealand as they are my firm favourite. Though much larger than the European variety, they are tender and extremely tasty. In England many supermarkets sell them frozen. If you can't find green-shell mussels, use a smaller variety – they'll just be less decorative! This dish is a real eye-catcher served in the individual mussel shells. (For photograph see page 54.)

SERVES SIX AS A STARTER
24 green-shell mussels, steamed open and removed from their shells
1 mild red chilli
30g (1¹/₄oz) dried hijiki seaweed, soaked in cold water for at least 3 hours
3 tablespoons sesame seeds, toasted
¹/₂ cup fresh coriander leaves
100ml (3¹/₂fl oz) extra virgin olive oil
2 tablespoons sesame oil
2 bunches rocket, leaves picked from the stalks and washed

PICKLED VEGETABLES
250ml (9fl oz) water
150ml (5fl oz) cider vinegar
2 whole star anise
100g (3¹/₂oz) caster sugar
1 12cm (5in) piece of cucumber, seeds taken out and finely sliced
1 12cm (5in) piece of mooli (daikon), peeled and julienned

You'll need to make your pickles a day ahead. Bring the water, vinegar, star anise and sugar to the boil. Mix the cucumber and mooli and put them into a heatproof jar that has been warmed with hot tap water. Pour in the pickling liquid (with the anise) and seal the jar. Allow to cool completely, then store in the fridge overnight.

The next day, slice the seeded chilli finely and add it to the mussels and the drained hijiki in a large bowl; take a quarter of the liquid from the pickles and add to the bowl. Leave to marinate for 2 hours. Drain the vegetables completely, discarding the star anise and liquid (or keep in the fridge for dressing salads), and add with the rest of the ingredients to the mussels. Serve immediately.

fish

grilled
mackerel
with a sweet chilli glaze
on mixed bean salad

FOR SIX

6 large boneless mackerel fillets

BEAN SALAD

A selection of mangetout, green beans, broad beans and sugar snaps
20ml lemon juice
2 pinches sea salt
60ml extra virgin olive oil

SWEET CHILLI GLAZE

1 teaspoon dried chilli flakes
1/2 cup light brown sugar
2 teaspoons light soy sauce
4 teaspoons lemon juice
1/4 teaspoon finely ground allspice

Prepare the bean salad by lightly blanching the beans in separate batches until they are just cooked but still have a crunch. Refresh them under cold water and dress with a little lemon juice, sea salt and extra virgin olive oil.

Put all the ingredients for the sweet chilli glaze into a small pan and bring to the boil, cook until the mix starts to thicken and remove from the heat. Place the mackerel fillets, skin side up, on an oiled piece of foil under a hot grill for 2 minutes, then turn over and grill for another minute. Cooking times will vary with the thickness of the fillet – aim to keep the fish pink in the middle. Brush with the glaze and return to the grill until the glaze just begins to 'burn'. Remove from the tray and serve on the beans.

Mackerel works particularly well with the sweet glaze and cool salad because of its oily content, though salmon fillets are also delicious. Mackerel is rarely seen on restaurant menus, mainly because of its reputation as a cheap and unadventurous fish. But it has fantastic taste and texture. Bring back the mackerel!

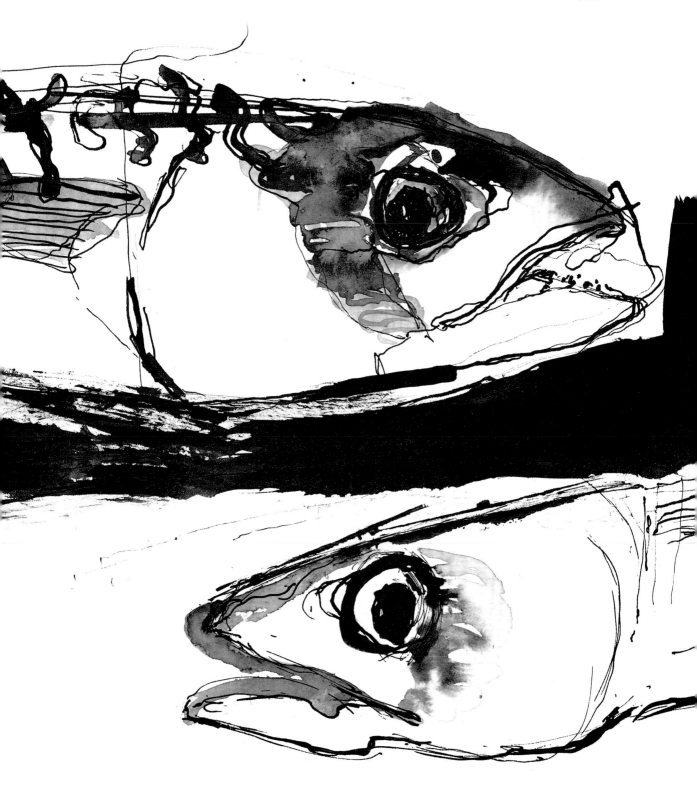

grilled squid
with chorizo,
tomato & new potatoes

FOR FOUR LARGE STARTERS

4 medium-sized squid, cleaned and separated into body and tentacles
150ml (5fl oz) olive oil
Salt and freshly ground black pepper
200g (7oz) cooking chorizo, preferably spicy, sliced on an angle into 8 pieces
2 ripe tomatoes, cut in half
8 new potatoes, scrubbed, boiled, and then cut into halves
1/2 cup fresh flat-leaf parsley leaves
50ml (1 3/4 fl oz) lemon juice

Squid should be cooked either hot and fast or slow and long – anywhere between will make it chewy and tough. For this dish have your grill, griddle or even a frying pan very hot before starting to cook. Holding the body of the squid, cut it in half lengthways from the pointed 'tail' end to the body cavity. Toss the flesh in a few teaspoons of the oil, some salt and pepper and then put under the grill; turn after 45 seconds and grill the other side. The heat will make the body curl up, but don't worry. Remove to a large warm plate and do the tentacles, which need about 30 seconds longer to cook than the body.

Next grill the chorizo pieces for about 30 seconds each side. Put them on the plate too and grill the tomatoes for about 1 minute each side. Mix the potato with the parsley and divide between four plates. Assemble the grilled foods on top, drizzling with some of the lemon juice and the remaining olive oil. Sprinkle with seasoning and eat before it gets too cold.

I love the whole concept of squid. It's an odd protein in that it's thought of as 'fish' – yet it isn't a fish at all. I suppose on a classical restaurant menu it would have to share a space with cuttlefish, that other seafood oddity. Part of the fun (if you can call it that!) of using squid is in the cleaning of it. Most people buy it already prepared, which saves a lot of mess and time, but one day try to do it yourself. Basically you take out all the inside mess, remove the beak from the tentacles and peel away the tough membrane that covers the whole beast. Easy! Just be sure to wear gloves. This dish has its roots in Spanish cooking.

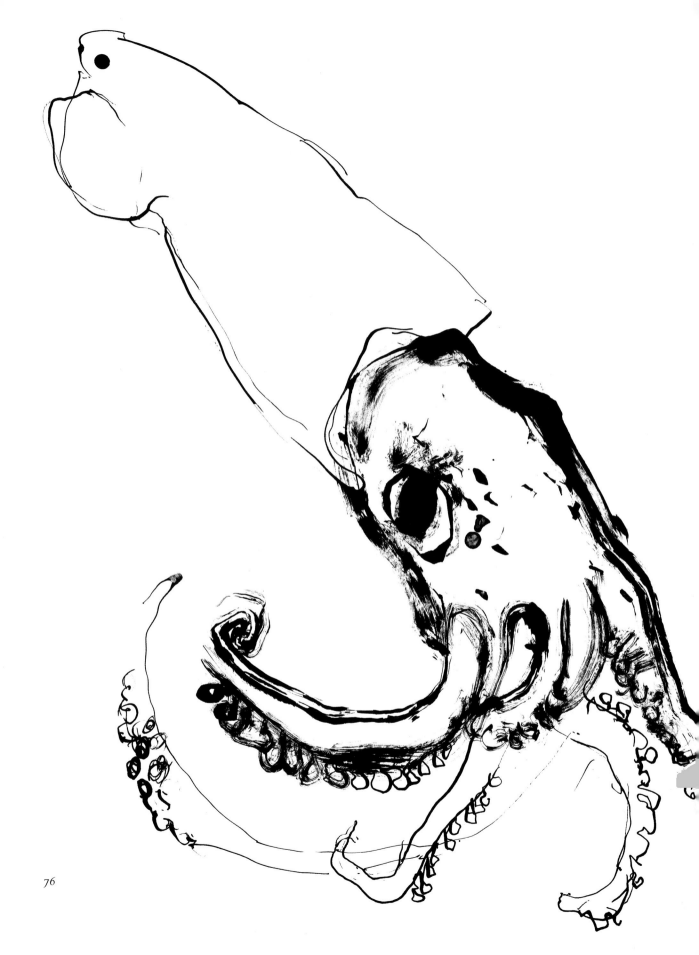

squid,
coriander & lime salad
with pickled vegetables

This is a perfect dish for summer and if you already have some pickles in your fridge it is quick to make up. The pickles give the dish a lovely crunch; the recipe given here makes enough to set some aside for future occasions.

PICKLED VEGETABLES

300ml (11 fl oz) cider vinegar
600ml (1 pint) water
200g (7 oz) caster sugar
1 teaspoon finely ground star anise
2 tablespoons grated peeled ginger
3 red peppers, cut in half and seeds removed
1 20cm (8 in) piece of mooli (daikon), peeled
4 medium-sized carrots, peeled
1 cucumber

SQUID (FOR FOUR STARTERS)

500g (18 oz) cleaned squid
  (medium size or small is best for this dish)
2 teaspoons sesame seeds
40ml (1 1/2 fl oz) sesame oil
120ml (4 fl oz) extra virgin olive oil
1 cup fresh coriander leaves
1 cup finely sliced spring onions
4 limes, juiced and the zest finely grated
1 tablespoon Asian fish sauce

Bring the vinegar, water, sugar, star anise and ginger to the boil, put a lid on the pot and simmer for 5 minutes. Meanwhile, finely julienne the peppers, mooli and carrots. Slice the cucumber into thin discs and mix in a large bowl with the other vegetables. Put these into a large glass or ceramic preserving jar and sit the jar in a tub of hot water for a few minutes to warm. Bring the pickling liquid back to the boil and pour over the vegetables, shake the jar gently to release air bubbles and seal. When cold, store in the fridge, turning over once a day. Leave for at least a week before using.

Slice the squid bodies into fine rings, divide the tentacles into two halves and keep the fins whole. Bring a pot of salted water to the boil. Meanwhile, fry the sesame seeds in the sesame oil until they turn golden brown. Pour this mix into a heatproof bowl and add the olive oil, coriander leaves, spring onions and lime juice and zest. When the water is boiling rapidly, carefully add all the squid to the pot and stir gently. After 40 seconds pour the squid into a colander and drain well. Add the squid – while still hot – to the marinade and mix well. Add the fish sauce and mix again, then leave to cool. Leave to sit for an hour before serving to allow the flavours to develop.

Divide the squid between four plates, serve a good handful of the vegetables on top and drizzle with some pickle liquid as a dressing.

tuna tartare
with roast tomatoes

Tuna, line-caught and dolphin-friendly of course, is a delicious treat. At home in London we tend to eat it raw with tamari and wasabi, or grill it rare, drizzle with a good lemony olive oil and serve it with potatoes and salad. The following is a dish I made for a dinner at home for my partner Michael when he requested a seafood buffet for his birthday. With everything cut a little smaller it also makes a good canapé on crunchy toast.

FOR SIX STARTERS

500g (18oz) tuna loin

6 tomatoes

Salt and pepper

100ml (3$^{1}/_{2}$fl oz) extra virgin olive oil

50ml (1$^{3}/_{4}$fl oz) balsamic vinegar

$^{1}/_{2}$ cup sliced spring onions

$^{1}/_{4}$ cup chopped fresh chives

$^{1}/_{2}$ cup sliced fresh basil

$^{1}/_{4}$ cup small fresh mint leaves

$^{1}/_{2}$ green chilli, seeds removed and julienned

2 tablespoons finely julienned peeled ginger

135ml (4$^{1}/_{2}$fl oz) lime or lemon juice

50ml (1$^{3}/_{4}$fl oz) tamari

Set the oven to 140°C/275°F/Gas 1. Cut the tomatoes into quarters, put in a ceramic dish, season with salt and pepper and drizzle with half of the olive oil and half of the balsamic vinegar. Bake in the oven for 1$^{1}/_{2}$ hours, then remove and drizzle with the remaining oil and vinegar.

Trim the tuna loin of any sinews, but don't worry about the black part of the flesh – it tastes very good! Cut into 1cm ($^{1}/_{3}$in) dice, mix with the next seven ingredients and leave to sit at room temperature for 1 hour. Just before serving, drain the olive oil and vinegar from the tomatoes, add along with the tamari to the tuna and mix in gently. Serve with the tomatoes.

pacific ceviche

My Gran first made this dish after one of her visits to Fiji and it became a family favourite. We lived near the coast, and in summer we'd be out at sea just after sunrise bringing in a catch. Then it was back to the house for a breakfast of fried fish before going off to school. Later, for long hot lunches around the pool, this ceviche was a great refresher. Any fish will do as long as it's skinned and boned.

FOR FOUR LARGE STARTERS
600g (1 1/4 lb) freshest fish fillets; snapper, hake, monkfish and
*   mackerel all work well*
250ml (9 fl oz) lemon or lime juice
*   (5 lemons or 8 juicy limes is about right)*
1 teaspoon sea salt
1/2 teaspoon finely chopped green chilli
400ml (14 fl oz) unsweetened coconut milk
1/4 teaspoon freshly ground white pepper
1/2 cucumber
1 cup watercress, cut into 1cm (1/3 in) lengths

Slice the fish into pieces 5mm (1/5 in) thick and put into a non-reactive bowl with three-quarters of the citrus juice, mix well and leave to sit in the fridge, covered, for 2 hours. This 'chemically' cooks the fish.

Drain the fish in a colander and return to a clean bowl. Add the salt, chilli, coconut milk, pepper and remaining juice. Mix well and return to the fridge. You can now leave this for up to 6 hours before finishing the dish.

Just before serving, peel and de-seed the cucumber and cut it into chunks. Add these and the watercress to the fish and mix well; test for seasoning and serve immediately.

Skate is a much underrated fish in Britain and one which is usually seen pan-fried with capers and burnt butter. Nothing wrong with that of course, but, well, it's just another fish. In fact it's as versatile as any but with the advantage of being quite different in texture and shape. There are two schools of thought on when to cook skate. One is to wait until it smells slightly of ammonia; the other, which I favour, is to use it as fresh as possible – within two days of being caught.

deep-fried
skate wing
with smoked paprika aïoli

Get your fishmonger to remove the skate's leathery outer skin as it's nigh impossible to do at home. Also, for this recipe, ask him to cut the flesh from the cartilage. You'll get two pieces from each wing, which you can then cook as you would any other fish. If you can't find smoked paprika, a mixture of cayenne pepper and regular paprika will do fine.

FOR SIX MAIN COURSES

6 300–350g (11–12 oz) skate wings, prepared as described above

1/2 cup plain flour for dusting the fish

Vegetable oil for deep-frying

3 limes

BEER BATTER

1 1/2 cups plain flour

2 teaspoons smoked paprika

1 teaspoon salt

2 teaspoons demerara sugar

1 teaspoon baking powder

230ml (8 fl oz) beer, at room temperature

SMOKED PAPRIKA AÏOLI

2 egg yolks

1 egg white

1/2 teaspoon salt

50ml (1 3/4 fl oz) lime juice

4 cloves of garlic, peeled

1 tablespoon seed mustard

3 teaspoons smoked paprika

400ml (14 fl oz) olive oil

Fill a deep-fryer with oil and leave to heat to the correct temperature. Meanwhile, take the dry ingredients for the batter and whisk in a bowl for a few seconds to mix them. Add all the beer and whisk it in, starting from the inside and working outwards. Make sure there are no lumps and leave the batter to sit for 15 minutes.

For the aïoli, put the egg yolks and white, salt, lime juice, garlic, mustard and smoked paprika into a small food processor and purée for 1 minute. Slowly drizzle in the olive oil (don't use extra virgin), making sure it is absorbed and the aïoli doesn't separate. If it does separate, transfer 100ml (3 1/2 fl oz) to another bowl, add one more egg yolk and whisk well. Once it has taken, whisk the remaining mixture in gradually.

Lightly dust the skate wings with flour and dip them, one by one, in the batter. Hold them above the batter and let the excess drip back into the mixture, then place them carefully into the deep-fryer. Cook only as many as can easily fit into your fryer at one time – if you overcrowd it the temperature will drop and the batter will not crisp. They should take 2–3 minutes to cook. Drain on absorbent paper and serve with a crisp, simple salad, such as wild rocket or watercress, and plain boiled new potatoes. Serve the aïoli and some fresh limes on the side.

roast turbot
marinated in spicy lime & coconut
on cucumber & coriander salad

This recipe is perfect with a fish like turbot, but it also works well with chicken suprêmes and lamb loins. In the marinade fresh lime peel can be used in place of the lime leaves and salt or tamari instead of the fish sauce. If you like things hot, include the seeds and pith of the chilli.

FOR FOUR

4 large pieces of turbot fillet, around 200g (7oz) each, skinned
Cooking oil

MARINADE	SALAD
400ml (14 fl oz) unsweetened coconut milk	*1 large cucumber*
8 lime leaves	*1/2 cup lemon juice*
1 whole lime, cut into eighths	*1/2 cup caster sugar*
1 red chilli	*1/2 teaspoon salt*
1 tablespoon Asian fish sauce	*1 cup fresh coriander leaves*
2 tablespoons grated fresh ginger	

Preheat the oven to 200°C/400°F/Gas 6. Purée the ingredients for the marinade in a blender for 1 minute. Pour over the fish in a non-reactive bowl and refrigerate for at least 12 hours, turning a few times. Peel the cucumber into ribbons (or slice it finely), discarding the seeds. Mix with the remaining ingredients – except the coriander – and keep in the fridge in a non-reactive bowl.

In the oven heat a shallow ovenproof dish in which you've put a few tablespoons of cooking oil. Once it's hot, carefully place the fish and the marinade in it, avoiding any splatters, and return to the oven; cook for 4–8 minutes (the exact time will depend on the thickness of the fillet). While it's cooking, drain most of the liquid from the cucumber, mix the coriander through it and serve under the fish. Plain boiled rice is an excellent accompaniment if you want a more filling meal.

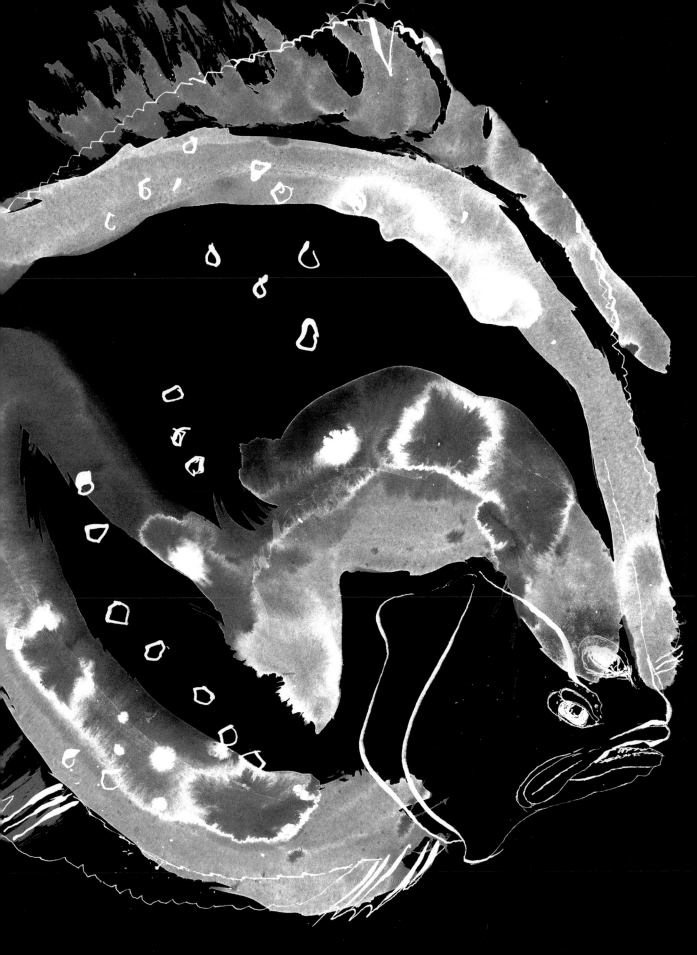

Apart from the obvious visual attraction of its combination of colours, this dish offers some exquisite tastes. The rice is slightly tangy, the curry sweet and spicy and the sauce fresh and clean. Monkfish is a good fish to use as it remains firm and retains its shape, but you could replace it with most other fish or even chicken. The amount of chilli used here will make a really fiery curry, so moderate it if you feel cautious!

monkfish
coconut curry
on red rice with green sauce

FOR SIX MAIN COURSES

6 cloves of garlic, peeled

1 6cm (2^1/$_2$ in) piece of fresh ginger, peeled

3 hot green chillies

6 lime leaves

2 teaspoons shrimp paste

1 teaspoon finely ground star anise

1 teaspoon fennel seeds, dry-toasted

100ml (3^1/$_2$ fl oz) sesame oil

2 medium red onions, peeled and finely diced

1.2kg (2^1/$_2$ lb) cleaned monkfish fillets,
   cut into 12 equal-sized pieces

600ml (1 pint) unsweetened coconut milk

4 tablespoons Asian fish sauce

RED RICE

2 cups basmati or Thai fragrant rice

6 tablespoons tomato paste

3^3/$_4$ cups cold water

GREEN SAUCE

1 cup fresh coriander leaves

1 cup fresh mint leaves

1 cup fresh basil leaves

2 bunches spring onions, trimmed

2 teaspoons finely grated lime zest

1 cup salad oil

Rinse the rice in a colander under running water for a minute to remove rice dust. Put it into a heavy saucepan with the tomato paste and the water and stir well. Place on a high heat until it comes to the boil, stirring twice during this time. Cover with a tight-fitting lid and turn to a very low simmer. Cook for 12 minutes before turning off the heat and removing to a warm place to continue cooking in its own heat. Leave for 15 minutes before removing the lid and serving.

Put the first seven ingredients of the fish recipe and half the sesame oil into a food processor and purée to a fine paste. Heat up a heavy saucepan and then add the remaining oil; put in the onion and fry over a high heat for 2 minutes, stirring well. Add the fish and fry for a minute until it is sealed all over. Remove the onions and fish, return the pan to a high heat and add the puréed paste. Fry for 1 minute, stirring well, then return the fish and onion to the pan. Add the coconut milk and bring to a rapid simmer; add the fish sauce and cook for at least 5 minutes until the fish is done.

To make the green sauce, simply put all its ingredients into a food processor and purée to a paste. Put the fish curry on top of the red rice, spoon the green sauce over them and serve.

baked haddock
with tomatoes, saffron & wine on mash

This is an ideal dish for those cool autumn nights when tomatoes are still in season and deliciously red and ripe. The mustard mash on page 129 goes well with it, but the mash can be made without mustard if you prefer. Cod is a good substitute for haddock. If your tomatoes are really ripe use a dry white wine; if they are a little underripe, go for a slightly sweeter one. A New Zealand chef, Anna Hansen, gave me the idea for the dish, which is simple to prepare and extremely tasty. (For photograph see page 70.)

FOR SIX

6 200g (7oz) pieces of haddock fillet, bones removed
Salt and pepper
4 tablespoons olive oil
6 tomatoes, halved and then cut into thin slices
$^1/_4$ teaspoon saffron, soaked in 100ml ($3^1/_2$ fl oz) white wine for 20 minutes
450ml (15 fl oz) white wine
4 cups hot mashed potatoes (see page 129)
4 tablespoons extra virgin olive oil

Season the fish with salt and freshly ground black pepper. Heat a large pan (or two smaller pans) that will hold all six fillets comfortably. Add the oil and when it's hot put the fish in, skin side down, and fry for 2 minutes. Turn the fish over carefully and add the tomatoes, saffron and wine, turn the heat up to full and cover the pan tightly with a lid or foil. Cooking time will depend on the thickness of the fillets, but check them after 4 minutes.

When they're done remove the fillets carefully, place each on top of a pile of hot mash and keep warm. With the sauce on full boil, add the olive oil and reduce for a minute or two until the sauce thickens, then spoon it over the fish and eat immediately.

blackened
salmon
with citrus labne

Blackened foods come from the southern states of America, while the labne is Middle Eastern and takes its name from the Arabic for 'white'. The two styles blend surprisingly well in this dish. I recommend using fillet from the thicker, head end of the fish, although a tail cut is quite acceptable. You'll have some blackening herbs left after making this, so use them to experiment with different meats and fish.

FOR SIX STARTERS

800g (1³/4 lb) salmon fillet, boned and skinned
2 tablespoons vegetable oil

BLACKENING MIX

¹/4 cup dried thyme
¹/4 cup dried ground rosemary
¹/4 cup dried oregano
¹/4 cup paprika
1 tablespoon garlic powder
1 tablespoon cumin seeds
2 tablespoons fine salt

CITRUS LABNE

300ml (11 fl oz) Greek yoghurt
1 lemon
1 lime
1 orange
50ml (1³/4 fl oz) extra virgin olive oil

For the citrus labne, line a colander with clean muslin or J-cloth, tip the yoghurt in and fold the cloth over the top. Place the colander on a soup bowl and leave in the fridge overnight to drain. Next day tip the drained yoghurt into a bowl and mix in the finely grated citrus peel and olive oil and, last, the juice from the lime. (The liquid that drains from the yoghurt can be used for a bread mix or thrown away.)

Mix all the herbs for the blackening mix together. Give the salmon a generous coating and leave to sit for 5 minutes. Get a heavy frying pan very hot, then put in the vegetable oil, put the salmon in immediately and cook for 90 seconds without moving it, flip and cook for another 90 seconds, then remove and leave in a warm place for 15 minutes. Halve the cooking times if you are using a tail cut.

Slice the salmon into 12 quite thin portions and serve with the labne and a crisp salad.

fowl

roast chicken
& sweet potatoes
with olive & basil sauce

FOR FOUR

4 orange-fleshed sweet potatoes

1 large chicken, jointed so the legs and breasts are cut into halves (i.e. 8 pieces in total)

Olive oil

Salt and freshly ground black pepper

OLIVE AND BASIL SAUCE

1 cup pitted green olives

250ml (9fl oz) dry white wine

2 tablespoons honey

2 cups fresh basil leaves

1 bay leaf

6 cloves of garlic, peeled

2 lemons – peel and juice

4 tablespoons light soy sauce

The oven should be preheated to 200°C/400°F/Gas 6. Scrub the sweet potatoes and slice them into 1cm (⅓in) thick pieces, brush lightly with olive oil and season and place in the bottom of a ceramic roasting dish. Brown the chicken pieces in olive oil, then place the legs only on top of the sweet potatoes and put in the oven for 25 minutes. Meanwhile, make the olive sauce by blending the ingredients in a blender or small food processor until fairly fine but not a purée.

When the 25 minutes are up, remove the dish from the oven, turn the legs over and add the breast pieces before returning to the oven and cooking for a further 20 minutes. Then take it out again, pour the sauce over the chicken and return to the oven until the chicken is cooked. To check, insert a fine knife into the thickest part of the breast: when the fluid runs out clear the meat is done. Serve with a good mixed salad.

This is one of those easy-to-knock-up-at-the-last-minute dishes that nevertheless taste exceedingly good. It's also delicious cold. Use a free-range bird or at least a corn-fed one. Sweet potatoes from the Pacific and olives from Europe may not seem natural soulmates but, well, eating is believing. It's that sweet and salty combination again – you'll find they work marvellously well together.

grilled cinnamon quail
on roast carrot & ginger salad with pomegranate

Pomegranate syrup is a sweet–sharp cordial that's sold in Persian food stores and many Middle Eastern shops. If you can't get hold of it, use tamarind paste or a light honey (with either of these halve the amount to 1 tablespoon). Quail is a tasty bird that's fun to eat with your fingers. They are easily roasted by brushing with oil and cooking in an oven at 200°C/400°F/Gas 6 for 12–15 minutes. In this recipe they're grilled, so flatten them out beforehand as they will then cook a lot quicker and more evenly. Holding the bird in one hand, breast side down, cut along both sides of the backbone and remove it. Then lay the quail on a board and flatten it with the palm of your hand. (For photograph see page 90.)

FOR SIX STARTERS

6 quail, flattened out as described

1 cinnamon stick, ground in a spice grinder
  (or 1 1/2 teaspoons powdered cinnamon)

2 tablespoons pomegranate syrup

3 teaspoons tamari

Grated zest of 1 lemon

80ml (3fl oz) olive oil

2 medium-sized carrots, peeled and finely sliced

2 medium-sized red onions, peeled and finely sliced

Grated zest and juice of 1 lemon

2 tablespoons grated peeled ginger

50ml (1 3/4 fl oz) extra virgin olive oil

1/2 teaspoon salt

1/2 teaspoon coarsely ground black pepper

1/2 cup fresh coriander leaves

Seeds of 1 very red and ripe pomegranate

Mix the cinnamon, pomegranate syrup, tamari, the first amount of lemon zest and the olive oil in a bowl. Add the prepared quails to this paste and mix well, then seal the bowl with clingfilm and leave to marinate for at least 4 hours.

While they are marinating make the salad. Turn the oven to 180°C/350°F/Gas 4. Mix all the remaining ingredients except the coriander leaves and pomegranate seeds and place in a roasting dish in the oven. Stir after 15 minutes and cook for a further 20 minutes. Keep an eye on them as they may burn; they will need more occasional stirring. Once cooked, remove the dish from the oven and when it's cooled mix in the coriander leaves and put to one side. The other thing to do at this stage is to extract the seeds from the pomegranate. Cut it into quarters and take them out, removing any clinging white membrane.

Turn the grill on full. Lightly oil a piece of foil and lay it on a baking tray that will fit comfortably under the grill. The tray should be about 7-10cm (3-4in) from the elements. Sit the birds on the foil, side up, and grill for 3½ minutes, then turn over and grill for 4 minutes on the other side. Because the syrup has a high sugar content it will caramelise, which is what we want. However, it's not as nice if it burns, so turn the heat down a bit if necessary. Quails can be eaten a little pink, so test by pushing a sharp knife into the thigh of a bird. If the juices run out clear or just a little pink they are cooked.

Serve the quail while still warm on top of the carrot salad and sprinkle with pomegranate seeds. If eating with fingers, remember to provide finger bowls!

poached guinea fowl
with broad bean, garlic, pear & fennel salad

This is a salad that can be served as either a main course or a starter. If the former, it should be eaten with roast new potatoes (or even mash) for a more filling meal. It can be served cold or the fowl, beans and garlic can be warmed in the poaching liquor and served on the pear salad. The flavours in this dish may seem to combine a bit of everything from all over the place but, trust me, it works. A succulent free-range chicken can be substituted for the more exotic guinea fowl.

FOR SIX MAIN COURSES

1 guinea fowl, about 1.6–2kg (3½–4½ lb)
1 head of garlic – pull the cloves apart but don't bother to peel
600ml (1 pint) dry white wine
2 bay leaves
1 10cm (4in) piece of rosemary
Small bunch of oregano
1 lemon, cut in half
1 leek, cut into 5 and washed
2 carrots, peeled
2 white onions, peeled and cut into quarters
4 medium heads of fennel (trim the tops off and keep to use in the poaching liquor)
Salt
3 cups podded broad beans (smaller are sweeter)
50ml (1¾fl oz) lemon juice
2 sweet pears

Remove any packets of giblets from the bird and wash the cavity well. Tie the legs together to make it easier to manoeuvre in the pot. Put the next nine ingredients plus the fennel trimmings into a large pot – one that will comfortably hold the whole bird – and bring to the boil. Add the bird and pour in enough cold water to cover it by 6cm (2½in). Return to the boil, then turn to a rapid simmer and add 2 tablespoons of salt. Put a lid on and poach for 1 hour (don't allow to boil as the flesh will toughen), turning the bird over after 30 minutes to ensure even cooking. After an hour push a sharp knife into the upper thigh – the meat should not be at all bloody or too pink. If not done, keep going. Turn the heat off and leave the bird to cool in the liquor.

At this point, ladle out 500ml (18fl oz) of the liquor and bring it to the boil in a pot. Add the broad beans and boil for 2–4 minutes (depending on their size) until a little firm to the teeth. When they're cooked, drain and put in a bowl of cold water, then squeeze the beans from the skins and leave to one side. Holding the fennel flat on its side, cut it as finely as possible into 'rings' and mix them with the lemon juice to prevent discolouring. Peel and quarter the pears, removing the cores, and slice into small wedges; mix with the lemon juice as well.

When the bird has cooled, take it from the pot and cut off the legs. Remove the skin and pull the flesh from the body and legs – a knife may help, but it should come away easily. Break or cut the flesh into smallish chunks, discarding sinews and bones. Add 1 cup of the poaching liquor, mix well and taste for seasoning (it'll probably need some salt). Remove the garlic cloves from the liquor, which can be strained and kept as a base for soups or laksas. Toss the garlic, broad beans, fennel and pear together and put on a plate. Place morsels of guinea fowl on top and serve.

guinea fowl
stewed with potatoes,
ginger, tamari & star anise

FOR FOUR

2 small guinea fowl

100ml (3¹/₂ fl oz) sesame oil

2 leeks, cut into 1cm (¹/₃ in) rings and washed well

20 new potatoes, scrubbed and washed

6 tablespoons coarsely grated fresh ginger

80ml (3 fl oz) tamari

8 star anise

10 cloves of garlic, peeled

1 8cm (3 in) piece of rosemary

1 litre (1³/₄ pints) chicken stock (or make it from the carcasses of the guinea fowl)

Salt and pepper

1 small lemon, cut into quarters

Take the guinea fowl, remove the legs and cut them at the knee joint to produce four pieces per bird. Remove the two breast bones, keeping the suprêmes on the bone. Lightly oil the portions with some of the sesame oil and fry to a golden brown in a frying pan. When they're done, transfer to a large pot.

Add the remaining oil to the frying pan and sauté the leeks for 5 minutes until they begin to colour, then add these to the pot. Now put all the other ingredients except the seasoning and the lemon in the pot as well and top up with stock, if needed, so that everything is covered. Put a lid on the pot and bring slowly to the boil before turning down and simmering for 1 hour. Skim off any scum that rises. Check for seasoning, then serve with the lemon quarters.

This is one of those dishes that lie somewhere between a soup and a stew, rather like the traditional *pot-au-feu*. It's easy to make and can be left to look after itself while you concentrate on other things. Bowls of the broth can be served as a first course, followed by the solid bits – with some crusty bread – for the main course. The broth can also be used to make a delicious polenta or risotto. It's an extremely versatile dish.

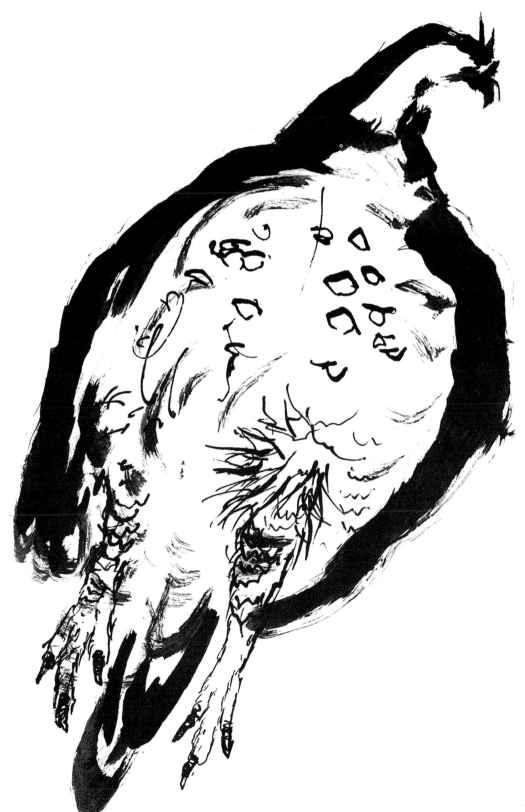

molly's
lemon roast chicken

Gran made this with kumera, the 'native' potato of New Zealand, which is a rich – almost purple – colour. In fact it came with the first canoes from Polynesia, but the New Zealand variety is different from what you'll find elsewhere in the Pacific. Kumera are unavailable in Britain, but it's becoming easier to get hold of sweet potatoes. The best ones are orange-fleshed and usually come from Israel or South Africa. In winter, parsnips make a great substitute. Gran's chicken is best eaten with a crunchy green salad.

FOR FOUR

1 large free-range or corn-fed chicken
Salt and pepper
150ml (5 fl oz) best olive oil
1kg (2¼lb) sweet potatoes
¼ cup fresh oregano leaves
2 teaspoons fresh rosemary leaves
2 medium-sized juicy lemons

Preheat the oven to 220°C/425°F/Gas 7. Remove the legs from the bird and cut them in two through the knee-joint. Remove the wings and the suprêmes and cut these into two also. (To make white stock for use on another occasion, put the carcass into a pot with some carrots, lemons, herbs and onions, cover with cold water, bring to the boil and simmer for 2 hours.)

Season the joints lightly with salt and freshly ground black pepper and brown in a little oil in a frying pan. Peel the sweet potatoes and cut into 2cm (¾in) dice, mix with the herbs, a little salt and pepper and half the oil and put in a ceramic roasting dish. Wipe the lemons and cut them in half lengthways, then slice finely. Place the leg joints on top of the potatoes, sprinkle with the lemon pieces and roast for 30 minutes. Add the suprêmes, drizzle the remaining oil on them and cook for a further 30 minutes. The chicken is cooked if the juices run clear when the flesh is pricked.

braised duck leg with duck liver
& roast garlic polenta

FOR SIX MAIN COURSES

6 large duck legs

18 cloves of garlic, peeled

1 carrot, peeled and diced

3 red onions, peeled and sliced

4 bay leaves

12 sage leaves

2 teaspoons fresh rosemary leaves

750ml red wine (standard bottle size)

Salt and pepper

1 litre (1 3/4 pints) chicken stock

250g (9oz) polenta grains, sieved to remove lumps

300g (11oz) duck livers, cleaned of sinew or fat and roughly chopped

1/2 cup roughly chopped fresh flat-leaf parsley

Set the oven at 170°C/340°F/Gas 3–4. Take a baking dish that's just large enough to hold the duck legs and the next seven ingredients. Put these in the dish and lay the legs on top. Season lightly and cover with foil, making a tight seal. Place on the middle shelf of the oven and cook for 2 hours. Check after an hour that the dish hasn't dried out; if it has, add some boiling water. When the 2 hours are up, take the foil off and remove everything from the dish except the legs. Return these to the oven for 10 minutes to brown.

Meanwhile, put the stock into a large 3 litre (5 pint) pot and bring it to a medium boil before slowly pouring in the polenta. Whisk firmly as you do this to avoid lumps. When all the polenta has been mixed in remove the whisk and stir in the livers and the ingredients from the roasting dish. Continue stirring over medium heat for 2 minutes and check for seasoning. At the last moment mix in the parsley. Serve the duck leg on a great dollop of polenta and with a finger bowl to the side to enjoy this meal to the full.

> This may seem an odd combination at first, but my guarantee that it's delicious is that we often eat it – and assorted variations – at home. It's a good way to savour two parts of the duck as well as soft polenta. Serve with steamed courgettes and beans.

pan-fried
partridge breast
on soba noodles
with horseradish & balsamic vinegar

Partridge is a delicious bird that I love to eat in the latter part of the year in London. They're wonderful roasted whole accompanied by sweet potato chips, roast garlic and bread sauce. But – as in this recipe – a good way to serve partridge is to take the breasts off the carcass and pan-fry them in butter and olive oil until they're just cooked. Any small bird can be treated in a similar way; just remember to cook the flesh quite rare and leave it to rest in a warm place for at least 5 minutes before eating. This allows the meat to relax and prevents that bleeding look that some cooked meat can have.

FOR SIX STARTERS

6 partridge breasts, off the carcass, all feathers removed and wing bones cut off
Salt and pepper
100g (3 1/2 oz) unsalted butter
30ml (1 fl oz) extra virgin olive oil
1 teaspoon fresh oregano leaves
2 tablespoons freshly grated horseradish
30ml (1 fl oz) good balsamic vinegar
100ml (3 1/2 fl oz) brown roast chicken stock (see page 164)
300g (11 oz) cooked soba noodles (see page 109)

Season the breasts lightly with salt and freshly cracked black pepper. Heat up a frying pan and, when it's hot, add the butter; let the butter melt and once it stops frothing add the olive oil; then put in the breasts, skin side down. Keep the heat on medium high and fry for 2 minutes before turning over and cooking for a further minute. Remove the breasts to a warmed plate and keep in a warm place.

Add the oregano to the pan and stir well, then add the horseradish, balsamic vinegar and stock and bring to the boil. Boil until the sauce has reduced by half, then take it off the heat and return the breasts to the pan. Heat up the soba noodles and serve the breasts on top with a covering of the sauce.

This is one of those dishes that I would dream about if I was stranded on a desert island. It is sweet, savoury, crisp and creamy and the aroma is heaven. It's easy to make and, once it's all cooking away, you can relax until it's ready. It reheats well, so it can be made in advance and kept for a few days in the fridge. To reheat, just put it into an ovenproof dish, cover with foil and warm at 160°C/325°F/Gas 3 for 30 minutes.

coconut &
duck leg curry
on coconut rice with pickled plums

The plums need to be made at least three days ahead, but they're best after ten. Covered, they will keep in the fridge for three months. There is no chilli in the recipe; add some if you want. To make the coconut rice, use the recipe for red rice on page 87 but leave out the tomato paste and replace half the water with unsweetened coconut milk.

FOR SIX MAIN COURSES

6 large duck legs, on the bone

450ml (3/4 pint) unsweetened coconut milk

50ml (1 3/4 fl oz) Asian fish sauce

12 cloves of garlic, unpeeled

3 cardamon pods, lightly crushed

2 stems of lemon grass, lightly bashed at the fat end

1/8 teaspoon saffron

3 onions, peeled and finely sliced

2 thumbs of ginger, peeled and finely julienned

1 teaspoon Sichuan peppercorns, lightly crushed

PICKLED PLUMS

1 kg (2 1/4 lb) very sweet ripe red plums, cut in half and stoned

1 cinnamon stick

4 star anise

1/2 red chilli

1/4 cup finely grated peeled ginger

500g (18 oz) unrefined caster sugar

1 teaspoon salt

1.3 litres (2 1/4 pints) water

700ml (24 fl oz) cider vinegar

Set the oven to 190°C/375°F/Gas 5. Put everything except the duck, coconut milk and fish sauce into a roasting dish large enough to hold the legs comfortably. Mix it all well and lay the legs on top. Now stir the coconut milk and fish sauce together and pour over the legs; the cooking liquid should come two-thirds of the way up the legs, so add some boiling water if necessary. Seal the dish tightly with foil and place in the top third of the oven. The legs should be cooked in 90–100 minutes – they are done when the flesh just begins to pull away from the bones. Take the foil off the dish after 90 minutes and keep cooking until the duck skin turns golden brown; the liquid should be bubbling, but it must not boil dry.

For the pickle, put everything except the plums into a pot and bring to the boil, then simmer for 5 minutes. Fill a large preserving jar with hot water and leave it until the 5 minutes is up. Tip the water out and pack the plums in, then pour the pickling liquid and its contents over the plums. If there isn't enough to cover the fruit, boil some water and vinegar (in a ratio of 2 to 1) and add to the jar. Seal immediately and leave to cool before placing in the fridge.

Serve the curry on top of the rice, with some pickled plums on the side.

diana stoll's
barbecued duck

FOR SIX

6 duck suprêmes
200ml (7 fl oz) balsamic vinegar
4 tablespoons drained capers
1 teaspoon sea salt
1/2 teaspoon freshly cracked black pepper
4 cloves of garlic, peeled and crushed

Marinate the duck two days in advance. Buy plump suprêmes off the bone and trim the fat to the degree you prefer (I love the taste of the crisped fat so I leave a good amount on). Lay them on a board, fat side up, and with a sharp knife score through to the flesh at 3mm (1/8in) intervals – the more you score, the crisper the skin will become. In a bowl mix the remaining ingredients and add the duck, turning several times. Put everything into a non-reactive container, cover, and keep in the fridge for two days. Turn the duck twice during this time.

On the day, get your barbecue (or grill) hot and at the same time heat up a frying pan. Remove the suprêmes from the marinade and reserve the capers in a few tablespoons of the marinade. Put the suprêmes – as many as will fit comfortably – skin side down into the pan and fry on a moderate to high heat without adding any fat until a light brown, draining off excess fat as you go. This can be done up to 6 hours before barbecuing, in which case cool them and store in the fridge.

When the barbecue is ready, set the rack 10cm (4in) above the charcoal, which should be red hot. Cook the suprêmes with the flesh side facing the heat for 5 minutes, then turn and cook for 3 minutes more (watch out for flare-ups). When they're done, remove from the heat and drizzle with capers and marinade, then rest for 10 minutes before handing out. Serve with crisp green vegetables or salad and some rice or boiled potatoes.

In June 1996 I was on Shelter Island, at the eastern end of Long Island, USA, helping to cook a wedding feast for one of my best friends. Diana Stoll had decided that she wanted barbecued duck suprême for her husband-to-be, Jeff Kinsel, and their 196 guests, so she hired a charcoal-burning monster of a grill. Here I pass on her recipe for this delicious dish. Be sure to score the fat really well and render as much of it from the suprêmes as possible before grilling as this will prevent flare-ups. Although best cooked on a barbecue, it can also be done under a domestic grill. Be warned, though, that as duck is extremely fatty a lot of smoke is inevitable.

red meat

lamb fillets
braised with tamarind & coconut
on black rice

In this dish the sweetness of the coconut, and indeed the lamb, is offset by the sharpness of the tamarind. I have served it both as a starter and, with lamb loin in place of the fillets, as a main course. Black rice is a native of Southeast Asia and is a real rice – it shouldn't be confused with Canadian wild rice, which is actually a form of grass seed. In Asia it is usually eaten as a dessert rice (as in the famous black rice and banana pudding of Ubud, Bali). If you can't find it, use organic brown rice as it's the nuttiness of the rice that works so well in this dish.

FOR SIX STARTERS

12 lamb fillets, each about 60g (2oz)
1 cup black rice
Salt
3 cups cold water
1 cup fresh coriander leaves
30ml (1 fl oz) sesame oil
400ml (14 fl oz) unsweetened coconut milk
60ml (2 fl oz) tamarind paste
2 tablespoons Asian fish sauce

Begin by cooking the rice. Rinse it well under cold water in a colander to remove dust and dirt and then put it into a deep, heavy-bottomed pot with half a teaspoon of salt and the three cups of water. Put a lid on and heat on an element turned to high. When it comes to the boil stir well and turn to a simmer, then put the lid back on. It will take a lot longer to cook than white rice, so after 15 minutes give it a gentle stir, add more water if needed and keep it cooking. Check again after another 10 minutes. This rice is very glutinous, so don't expect it to go fluffy – when cooked it will still be on the chewy side. Stir in half the coriander and put the rice to one side with the lid on.

Toss the lamb in the sesame oil and bring a frying pan to a high heat. Add all the fillets and brown on all sides. Pour in the coconut milk, tamarind paste and fish sauce and bring to the boil, then turn to a simmer and cook for 1 minute. Now remove the fillets – which should be nicely pink on the inside – to a warm plate. Turn the tamarind mixture up to a full boil again and reduce it by half.

Serve the lamb on top of the rice with the sauce poured over and the remaining coriander sprinkled on top of that.

sugar-cured beef
on ginger buckwheat noodles

This dish features a curing method I first learnt at Rogalsky's restaurant in Melbourne in the early 1980s. The flavours and ratio of sugar to salt vary with the type of meat that's being cured, as does the time of curing. In this recipe the beef goes a deep red in the middle and dark brown on the outside. The method became popular at the New Zealand Sugar Club when I cured hapuka (a New Zealand groper) – a fat-fleshed, tasty fish. In London I have successfully used tuna and to a lesser extent swordfish. Salmon also works well, particularly if you replace the ginger, star anise and tamari with juniper and tarragon.

FOR SIX STARTERS

500g (18oz) beef fillet, preferably from the centre
   of the fillet, trimmed of all fat and sinews
1kg (2¼lb) demerara sugar
700g (1lb 9oz) coarse sea salt
½ cup finely chopped unpeeled ginger
4 cloves of garlic, unpeeled and chopped
8 star anise, ground in a spice grinder
150ml (5fl oz) tamari
150ml (5fl oz) sesame oil

NOODLES

300g (11oz) buckwheat noodles (use the 40% buckwheat
   variety for this dish as they are more pliable)
2 tablespoons finely grated peeled ginger
2 tablespoons sake
2 tablespoons mirin
2 tablespoons lemon juice
1 tablespoon sesame oil
4 tablespoons peanut oil
½ cup finely sliced spring onions
½ cup chives cut into 2cm (¾in) lengths

To cure the beef, mix the first set of ingredients from the sugar to the oil together and place 1cm (⅓in) of the mix in the bottom of a rectangular dish that's just bigger than the fillet. Put the beef on this and spoon the remaining mixture over the top; cover the dish and put in the fridge. Turn the meat every 12 hours and remove from the fridge after 60 hours. Rub off excess mix and place on a cake rack to drain for 2 hours. The beef is now ready to use and, wrapped in greaseproof paper, will keep in the fridge for a week.

A macrobiotic friend showed me how to cook buckwheat (or soba) noodles, and it's a method well worth remembering if you find they always boil over. Bring 2 litres (3½ pints) of water to the boil in a 4 litre pan. Throw in the noodles and bring to a rapid boil, then pour in 200ml (7fl oz) cold water and bring back to the boil; pour in another 200ml cold water and this time when the noodles return to the boil drain in a colander, rinse gently with cold water and drain well.

Put the cold noodles into a big bowl and mix with all the remaining ingredients. Make tall mounds of the noodle salad on serving plates and put slices of beef on top.

smoked venison
with roast jerusalem artichokes,
pickled onions & smoked chilli

I first remember smoking meat and fish with my father Bruce as a child in Wanganui. Dad would build a metal box with a removable lid. Across the rim of the box there were metal rods from which he suspended the fish on hooks. Then we lit a fire and, when it was a mass of embers, we'd get it really smoky with a combination of branches and water to stop it flaring up. After a couple of hours the fish would emerge moist, smoky and delicious.

A version of the method is given with this recipe. It makes a lot of smoke, so attempt it only at an outdoor barbecue or if you have a powerful extractor. Smoking works well with almost anything – salmon, mackerel, chicken, garlic, chillies and tomatoes (combine the last three in a delicious salsa or soup). Here I use venison fillet, which is lean but rich in flavour. Beef and lamb fillets smoke just as well and are a little easier to get hold of. This is a really tasty dish, so I usually serve it as a starter. Smoke a minimum of 500g (18 oz) of meat as the effort required to set up your smoker will seem wasted on less. Plain tea gives the best results, but try other things smoked over jasmine or Earl Grey tea.

FOR SIX STARTERS

500g (18oz) venison fillet, trimmed of all fat and sinew

1 star anise, finely ground

2 juniper berries, crushed

$^{1}/_{4}$ cup roughly chopped fresh oregano

200g (7oz) coarse sea salt

200g (7oz) demerara sugar

150ml (5fl oz) sesame oil

1 red chilli, cut in half and seeds removed

$2^{1}/_{2}$ cups tea leaves

$1^{1}/_{2}$ cups white rice

500g (18oz) Jerusalem artichokes

50ml ($1^{3}/_{4}$fl oz) olive oil

2 medium-sized red onions, peeled and very finely sliced into rings

100ml ($3^{1}/_{2}$fl oz) lemon juice

30g (1oz) unrefined caster sugar

Salt and pepper

Extra virgin olive oil

Mix the star anise, juniper berries, oregano, sea salt, demerara sugar, sesame oil and the chilli together in a bowl. Add the venison and rub the mixture over it. Cover the bowl and turn the meat every half hour for 4 hours.

To make the smoker, get two metal roasting dishes and a cake rack slightly larger than the piece of venison – one, if not both, of the dishes must be deeper than the height of the fillet to ensure that it will be completely enveloped in smoke. Mix the tea leaves and rice together, tip them into one of the roasting dishes and lay the rack above this. If the rack is smaller than the dishes, you'll need to suspend it at least 6cm (2½in) above the bottom to prevent the meat from burning. This can be done by seating the rack on four metal dariole moulds.

Remove the venison from the marinade and lay it on the rack with the two chilli halves, then spoon quarter of a cup of the marinade over the meat, letting it dribble on to the tea mixture below. Invert the second roasting dish over the first and seal the join tightly with aluminium foil. Turn a cooking ring up to full and sit the 'smoking dish' on it. After 3–5 minutes smoke will come out – this is exactly what you want. Keep it smoking for another 8 minutes, then take it off the heat and pull off the foil. Remove the top roasting dish and check how far the venison has smoked. It should remain rare in the middle so that it doesn't become too dry. If you want it done more, reseal and return to the heat. When it's finished, remove the top roasting dish and leave the meat (with the chilli) on the rack to cool. Leave it to sit for 2 hours before carving.

Meanwhile, mix the onion slices, lemon juice, sugar and half a teaspoon of salt together well and leave to sit for at least 90 minutes to pickle lightly.

Set the oven to 200°C/400°F/Gas 6. Scrub the Jerusalem artichokes clean and slice them into 5mm (⅕in) pieces. Toss with the olive oil and some salt and pepper, then lay them on a baking tray and roast until cooked. Test by inserting a sharp knife as you would for a potato. Dice the smoked chilli finely.

To serve, carve the venison into 5mm (⅕in) slices, place on the roasted Jerusalem artichokes and top with some of the pickled onions. Complete the dish by sprinkling with the smoked chilli before drizzling with some of the pickling liquid and extra virgin olive oil.

kangaroo tail & olive stew
on soft roast garlic polenta

Now you may wonder where you will get kangaroo tail in England. Hopefully, with the growing interest in this fantastic meat you will be able to find it fairly easily. If not, use oxtail, another favourite of mine.

FOR SIX

2.5kg (5¹/2 lb) kangaroo tail cut through the bone

100ml (3¹/2 fl oz) vegetable oil

300g (11oz) carrots, peeled and cut into 1cm (¹/3 in) dice

300g (11oz) parsnips, peeled and cut into 1cm (¹/3 in) dice

500g (18oz) red onions, peeled and thickly sliced

300g (11oz) butternut squash,
*   peeled and cut into 1cm (¹/3 in) dice*

2 heads of garlic, peeled

1 litre (1³/4 pints) brown roast chicken stock (see page 164)

¹/2 cup fresh sage leaves

400ml (14 fl oz) best green olives (unpitted)

¹/2 cup Asian fish sauce

¹/4 cup soya sauce

500ml (18 fl oz) diced chopped tomatoes

¹/2 cup fresh thyme leaves

¹/2 cup fresh oregano leaves

Salt and pepper

SOFT ROAST GARLIC POLENTA

250g (9oz) polenta grains, sieved
*   to remove lumps*

1 head of garlic – roast whole in the oven
*   for 35 minutes, then peel and crush*

500ml (18 fl oz) light red wine

500ml (18 fl oz) brown roast chicken stock
*   (see page 164)*

50ml (1³/4 fl oz) tamari

1 teaspoon coarsely ground black pepper

1 cup sliced basil

Heat a deep braising pan on a ring, pour in the vegetable oil and, when it begins to smoke, add the kangaroo tail in batches and brown all over. Remove the tail and add the vegetables and garlic. Sauté for 5 minutes until they begin to brown, then add 250ml (9fl oz) of the stock and bring to a rapid boil, scraping the bottom of the pan to free goodies stuck there. Put the tail back in and add the ingredients from the sage to the tomatoes; return to the boil before reducing to a simmer and putting the lid on. The stew can be cooked either on a ring or in the oven at 170°C/340°F/Gas 3-4 for 2-2¹/2 hours. Keep the meat covered with liquid and skim off any fat after the first half hour. At the end stir in the thyme and oregano, and check the seasoning. Allow to rest for 20 minutes.

Now make the polenta. Bring the wine and stock to the boil. Pour the polenta into the simmering liquid in a steady stream, whisking as you go. When it returns to the boil stir in the garlic, tamari, pepper and basil and simmer for 5 minutes.

At home I serve the polenta and stew in two large, separate bowls with plenty of oven-toasted sweet potato bread (see page 199). If this sounds appealing, prepare the bread by breaking into uneven chunks and drizzle with lots of good olive oil, salt and pepper before toasting in the oven at 200°C/390°F/Gas 6 for 15 minutes.

spicy grilled
pork belly
on chinese greens

I think it's the fattiness of pork belly that puts a lot of people off. For me that fattiness is a compelling reason to eat it – maybe not a healthy one, but it is so delicious! The following recipe is a great way to enhance the flavour further with a sweet and spicy cooking method, which is Chinese in origin. For the Chinese greens, look at the recipe on page 143 and cook them with or without the shiitake mushrooms. The pork is also excellent with plain or coconut rice or on noodles. Another variation is to slice it thinly and add to a spicy laksa. If you make this in winter, tinned tomatoes can be used in place of fresh.

FOR SIX MAIN COURSES

1.8kg (4 lb) pork belly (ask the butcher to remove bones, but keep the rind on)

600g (1¼ lb) very ripe tomatoes

2 hot red chillies, stalks removed

2 tablespoons freshly ground star anise

2 tablespoons freshly ground cinnamon

4 cloves, ground

2 thumbs of ginger, roughly chopped

6 cloves of garlic, peeled

30ml (1 fl oz) Asian fish sauce

20ml (¾ fl oz) tamari

100g (3½ oz) demerara sugar

Using a very sharp knife, score the rind of the pork lengthways at 1cm (⅓in) intervals – cut into the fat but don't go as deep as the flesh. Put the remaining ingredients into a blender and purée to a fine paste. Get a dish that's just large enough to hold the pork and pour half the paste in, lay the pork belly on top and then pour on the remaining paste. Rub the paste into the flesh, especially into the score marks on the rind. Cover and leave for 1 hour.

Turn the oven to 180°C/350°F/Gas 4. Lay some foil in a shallow-sided baking tray and sit a cake rack on it – the rack must be larger than the pork belly. Put a baking shelf in the top third of the oven. Take the pork and lay it, rind side down, on the cake rack and place it in the oven. After 30 minutes baste it with some of the marinade left in the marinating dish. Roast for another hour, basting every 15 minutes. Then turn the belly over and cook for 30 minutes more. Test by cutting through the flesh a few centimetres from the end with a sharp knife. If it's a little pink, that's fine. The total cooking time will depend on the thickness of the pork – a particularly thick piece may need longer. If at any point it starts to burn, cover with foil but keep it roasting.

Leave the meat to go cold, then slice into 1cm (⅓in) thick pieces as you would bacon. Grill them over a barbecue or skillet or under a grill. Because the belly has a high fat content it may give off a lot of smoke and spit a little, so keep your eye on it and don't let it burn. Cook for 2 minutes on each side and serve on top of the greens.

beef pesto

If grilled scallops with sweet chilli sauce is the dish that's become our hallmark at the London Sugar Club, then it is beef pesto that was the star turn at the Wellington Sugar Club. When we opened in London we had New Zealanders phoning to ask if we were the same people that cooked beef pesto in Wellington and could we put it on the menu for their birthday, wedding or anniversary. The recipe may sound a little odd, as I'm told so much of my food does, but it really works. Make the pesto from the recipe on page 165. (For photograph see page 106.)

FOR SIX

1 piece of mid beef fillet, trimmed, about 1.2-1.5kg (2^1/$_2$ - 3^1/$_2$ lb)
500ml (18 fl oz) tamari
250ml (9 fl oz) cider vinegar
1 red chilli, moderately hot
12 cloves of garlic, peeled
^{1}/$_4$ cup seed mustard
150ml (5 fl oz) cider vinegar
^{1}/$_2$ teaspoon salt
^{1}/$_2$ teaspoon cracked black pepper
350ml (12 fl oz) olive oil
^{1}/$_2$ bunch chard, about 400g (14 oz) (in New Zealand this is called silver beet),
*   shredded finely, stems and all, and washed well to remove dirt*
3 courgettes, julienned
2 medium-sized beetroot, raw, peeled and finely julienned
1 cup pesto (see page 165)

Marinate the beef at least two days before you eat this. Put the tamari, the first amount of cider vinegar, the chilli and six cloves of the garlic into a blender and purée to a fine consistency for 30 seconds. Lay the fillet in a long dish, just large enough to hold it, and pour the marinade over it, cover with clingfilm and place in the fridge. Every 12 hours turn the beef over to expose all of it to the marinade. It can be left to marinate for up to a week.

Just before you cook the beef, take it out of the marinade and drain well, then dry with a cloth. Cut it into six equal pieces and leave to sit until you're ready to cook it.

Put the remaining garlic, the seed mustard, the second amount of cider vinegar, salt, pepper and olive oil into a food processor, purée for 30 seconds and pour into a large bowl. Bring a large pot of salted water to the boil, add the chard and stir well. After 30 seconds add the courgettes and stir, and after 1 minute drain it all through a colander. (The chard and courgettes can be steamed if you prefer.) Tip the hot vegetables into the bowl with the garlic dressing and stir well, add the beetroot and stir that in, then leave the bowl in a warm place.

Now heat up a grilling skillet or a grill to a high heat. Lightly oil the fillet on the cut sides and, for the best flavour, grill for no more than 2 minutes on each side. As it's marinated the beef won't need a lot of cooking, but of course if you like your meat well done, then cook it so. Just remember that the more you cook the more you will lose the exquisite succulence and flavour of this dish.

Serve the grilled beef on top of the warm chard salad and dollop the pesto on top of that. You will love it!

braised lamb shanks with moroccan spices
on parsnip couscous with harissa

In January 1995 two friends, Piers Thompson and Tanya Backhouse, got married. For the wedding dinner they requested that I cook this dish, as it was over this particular meal that Piers proposed to Tanya at a Mayfair club where I was then chef. So, in a tent in the grounds of Winchester College, in the middle of winter, I and three chef friends prepared the following recipe for 330 people. Stored in the fridge, the roast spices are useful for all sorts of things – from flavouring bread doughs or adding to roast chicken pieces to mixing into pasta with spinach and feta or adding to soups. The harissa is merely one of hundreds of variations, the simplest being puréed chillies, salt and vinegar. It too is versatile and can be stored in the fridge for up to two weeks.

FOR SIX PORTIONS

6 lamb shanks

4 tablespoons coriander seeds

4 tablespoons cumin seeds

4 tablespoons fennel seeds

6 whole star anise

4 cinnamon sticks

3 carrots, peeled and cut into 1cm ($^1/_3$in) dice

2 parsnips, peeled and cut into 1cm ($^1/_3$in) dice

6 small red onions, peeled and quartered

3 hot red chillies, sliced

150ml (5fl oz) cooking oil

3 juicy lemons, cut into 5mm ($^1/_5$in) dice

1 litre (1$^3/_4$ pints) chopped peeled plum tomatoes
*   (tinned can be used)*

3 teaspoons dried mint

1 cup unpitted olives – both green and black are fine

$^1/_2$ cup tamarind paste

100ml (3$^1/_2$fl oz) Asian fish sauce

50ml (1$^3/_4$fl oz) tamari

COUSCOUS

500g (18oz) couscous

3 medium parsnips, peeled and
*   cut into 1cm ($^1/_3$in) dice*

Salt

1 cup fresh coriander leaves

2 lemons – grate the zest, then juice

100ml (3$^1/_2$fl oz) extra virgin olive oil

HARISSA

Roast spices (see recipe)

4 red peppers

3 tablespoons dried mint

6 cloves of garlic, peeled

3 red chillies, stems removed and
*   cut in halves*

3 limes – grate the zest and then juice

200ml (7fl oz) extra virgin olive oil

1$^1/_2$ teaspoons sea salt

You can either cook the shanks in a heavy saucepan on top of the stove or braise them in the oven. If the latter, you'll need a braising dish that is deeper than the shanks to ensure they are covered with liquid as they cook. With the first method you'll need to roast them first anyway, so start by preheating the oven to 220°C/425°F/Gas 7. Put the shanks in a braising dish and roast for 40 minutes, turning over halfway through. (Save time by roasting the peppers for the harissa now – see the recipe in the next paragraph.) While they are browning, put the seeds, star anise and cinnamon on a baking tray and roast in the oven to a deep brown, taking care not to burn them. Remove from the oven and cool, then grind finely in a spice grinder, reserving half for the harissa.

Make the harissa while the lamb shanks are browning (or prepare it in advance). Add the peppers to the lamb pan and cook for 40 minutes, turning over halfway through. Remove from the oven and place in a plastic bag to cool. When cool, peel and remove the seeds. Put the flesh in a food processor and add the remaining ingredients, purée to a paste and it's ready.

Sauté the vegetables and chilli in the cooking oil until they soften and go slightly golden, then add the lemon dice and sauté for a further minute. Now add the tomato, roast spices, dried mint, olives, tamarind, fish sauce and tamari and bring to a gentle boil. Turn off and wait until the lamb shanks are ready.

When the lamb is ready drain off any fat in the roasting dish, then pour the tomato mixture on top and add enough warm water to come three-quarters of the way up the shanks. Seal the dish with foil and return to the oven for 2 hours, turning it down to 190°C/375°F/ Gas 5. After 1½ hours check that the liquid hasn't reduced too much; if it has, top it up with hot water. While the meat is cooking make the couscous.

Make the parsnip couscous 1 hour before eating. I use the couscous that most supermarkets and delis stock – it's precooked and just needs soaking. Soak in cold water and place the covered bowl in a warm place as this gives a much lighter and fluffier result than adding hot water from the start. Put the parsnips in a pot and cover with cold water. Add a teaspoon of salt and boil until cooked, then leave to cool for 5 minutes. Mix the coriander and lemon juice and zest with the oil. Put the couscous into a heatproof bowl, pour on the parsnips and their liquid and mix well. Add the oil and lemon mixture and mix again, then add enough cold water to cover it all by 5mm (⅕in). Mix well again, cover with clingfilm and stand in a warm place for at least 45 minutes.

Serve the shanks on top of a mound of couscous with lots of the tomato stew and have the harissa in a separate bowl so people can add as much as they want.

red lamb & pumpkin
coconut curry

As with anything spicy the amount of chilli to use depends on your taste. At home I tend to dish up curry at a medium heat and put out a bowl of fresh chillies, finely sliced and marinating in lime or lemon juice, so people can adjust it for themselves. Chilli is something that you soon grow accustomed to and crave more of. When I first arrived in Bali in 1985 I had only a slight taste for it from the Tabasco we used to have on rock oysters as children. By the time I reached Jogjakarta in Java two months later I was a devoted fan and have been ever since.

FOR SIX

100ml (3$^{1}/_{2}$ fl oz) sesame oil

1.5kg (3$^{1}/_{4}$ lb) leg of lamb, trimmed
   of fat and cut into 3cm (1in) dice

6 large red onions, peeled and sliced

3 carrots, peeled and diced

4 leeks, peeled, sliced and washed

$^{1}/_{2}$ cup finely diced ginger

10 lime leaves

10 cloves of garlic, peeled and halved

750ml (26fl oz) unsweetened coconut cream

750ml (26fl oz) brown roast chicken stock
   (see page 164)

1kg (2$^{1}/_{4}$ lb) peeled pumpkin,
   cut into 2cm ($^{3}/_{4}$ in) dice

RED CURRY PASTE

2 tablespoons blachan (shrimp paste)

12 cloves of garlic, peeled

6 tablespoons grated ginger

3 tablespoons grated galangal

3 lemon-grass stems, the outer two layers and top
   third discarded, the remainder peeled and
   coarsely chopped

10 lime leaves, chopped

10 red chillies

3 cinnamon sticks, freshly ground

8 cardamon pods, seeds freshly and finely ground

1 cup tomato paste

1 cup chopped fresh coriander stems and roots

To make the red curry paste, put the blachan in a metal pan and place under a grill for 1 minute – keep an eye on it and make sure that it doesn't burn. Then put it and the remaining ingredients for the curry paste into a food processor and blend to a fine paste. (This will make more than you need.)

In a large, deep saucepan heat the sesame oil and brown the lamb evenly on all sides; remove the meat and add the ingredients in the first list from the onions to the garlic and brown them too. Add half of the red curry paste and fry until aromatic. Return the meat to the pan and add the coconut cream and stock, bring to the boil and simmer gently for 1¾ hours, then add the pumpkin and cook till tender.

I would serve this dish with simple cooked Thai scented rice accompanied by some grilled flat bread and wok-fried bok choy.

thai-style salad of
lamb & peanuts
with lime dressing

FOR SIX STARTERS

6 limes

2 teaspoons coconut-palm sugar (demerara will do)

1/2 green chilli, seeds removed and finely sliced

1/2 teaspoon Asian fish sauce

1/2 teaspoon tamari

400g (14oz) lamb fillet, trimmed and cut into 6 equal strips

Vegetable oil

2 teaspoons rice, dry-roasted in the oven until golden and then finely ground in a spice mill

4 shallots, peeled and finely sliced

1 cup fresh coriander leaves

1/2 cup fresh mint leaves picked from the tips of the plants

1/2 cup roasted peanuts, coarsely ground

Grate the zest from three of the limes and then juice all of them; mix with the sugar, chilli, fish sauce and tamari until the sugar has dissolved. Sear the lamb strips in a very hot pan with a little oil, being careful not to cook beyond rare, and rest for a couple of minutes in a warm place. Then slice the lamb thinly and place in a mixing bowl. Add the dressing, roasted rice, shallots, herbs and peanuts and mix well. Divide into six mounds and eat immediately.

The first time I ate this salad (at a Thai restaurant in Brunswick Street, Melbourne) it was made with beef and I was blown away. A couple of years later I was discovering different versions of it in Thailand itself – so different that the beef was often replaced by chicken, duck or even green papayas! At home we add loads more coriander and, if you can get your hands on some Vietnamese mint, throw that in too.

roast lamb chump
on crunchy polenta with butter beans

Chump of lamb is just restaurant-speak for lamb rump. We generally get it trimmed up with a small square of fat left on top. The meat is very tender and juicy and usually 250–300g (9–11oz) in weight. The garnish in this recipe works well with any cut of lamb from grilled cutlets to roast leg. I prefer to eat lamb on the medium side of medium rare, but cook it however you like. Buy the best-quality dried butter beans as you'd be surprised how the quality of dried beans can differ.

FOR SIX MAIN COURSES

6 trimmed lamb chumps
300g (11oz) dried butter beans, soaked in cold water overnight, then drained
2 medium-sized leeks, finely sliced and rinsed clean
1/4 cup fresh rosemary leaves
10 cloves of garlic, peeled and roughly chopped
1/2 cup extra virgin olive oil
Salt and pepper
200g (7oz) polenta grains, sieved to remove lumps
Olive oil
1 medium-sized red onion, peeled and finely diced
12 fresh sage leaves, roughly chopped

Put the butter beans in a large pot and cover them with three times the amount of cold water. Bring to the boil, skimming off any foam, and turn to a rapid simmer. Add the sliced leeks, rosemary, half the garlic and the extra virgin olive oil. Stir well and cook for 1 hour, then add 300ml (11fl oz) cold water and continue cooking until the beans are tender. Make sure they are always covered with liquid and always season after cooking to stop the skins toughening.

While they're cooking make the polenta. Heat 100ml (3½fl oz) of the olive oil in a deep pot and fry the remaining garlic, onion and sage leaves over moderate heat for 6 minutes, stirring occasionally. Add 400ml (14fl oz) cold water, bring to the boil, and then add 1 teaspoon salt. With the water on a rapid simmer, pour in the polenta slowly in a continuous stream, whisking to prevent lumps from forming. Once it is all mixed in remove the whisk and stir with a wooden spoon for 1 minute. Oil a 20cm (8in) square baking dish, pour in the polenta and leave to cool.

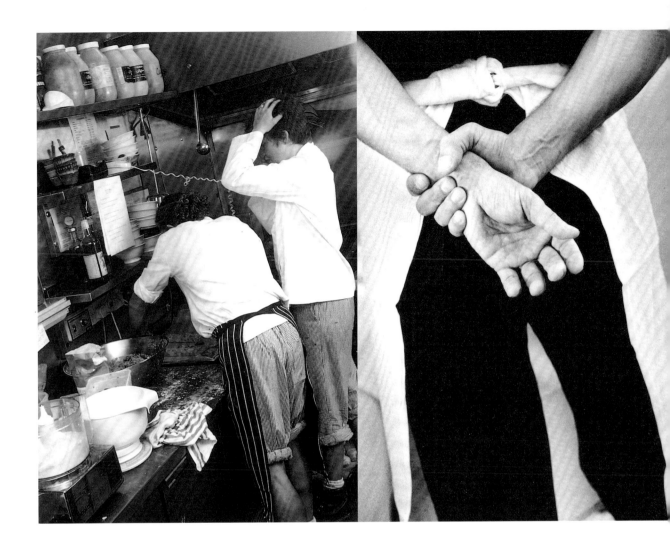

Turn the oven to 220°C/425°F/Gas 7. Cut the polenta into finger-sized pieces and brush with a little olive oil; lay these on a baking sheet and cook in the oven for 20 minutes.

Heat a frying pan up to very hot, then season the chumps with a little salt and pepper and brush with a little olive oil. Place them in the hot pan fat side down and fry for 4 minutes before turning over and frying for 3 minutes. Turn each chump on all four sides and fry for a minute on each, then place them in an oven dish that has been heating up in the oven. Cooking time will depend on the size of the chumps and how you prefer to eat them. I suggest 8 minutes. Then take them from the oven and leave in a warm place for 10 minutes before eating – this allows the juices to stay in the meat and the flesh to relax.

Serve the polenta fingers on the beans with the lamb sliced on top. Drizzle with some meat stock.

vegetables

The sweetcorn fritters are good served as a canapé – with crème fraîche, smoked salmon and chives on top – or made into larger fritters and topped with a salad of chopped rocket and tomato as a starter. I've even made them as a side dish for grilled chicken. The mix will keep in the fridge for a day; just remember to stir it well before cooking. The fritters can be eaten hot, cold or at room temperature.

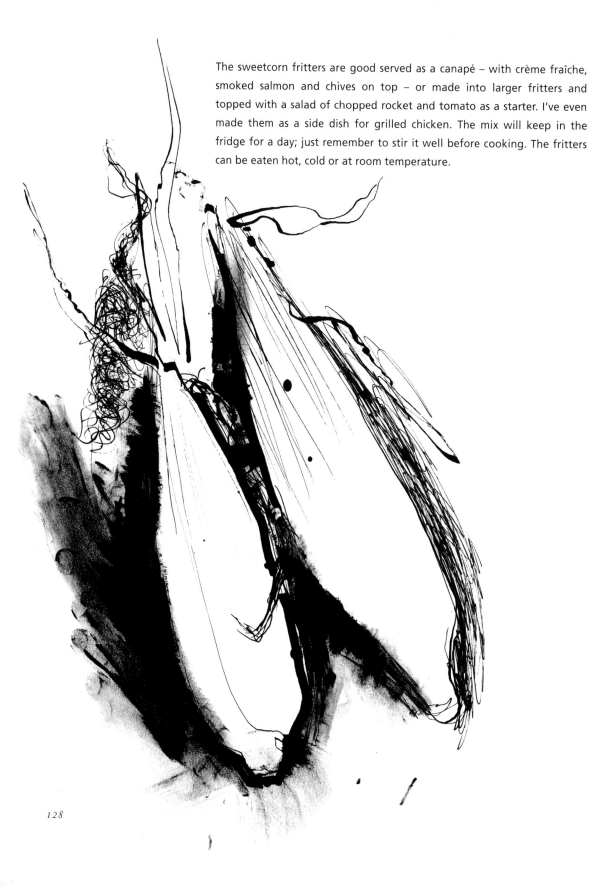

sweetcorn fritters

FOR ABOUT 15 SMALL FRITTERS

4 corn-cobs

4 eggs

$1/2$ cup crème fraîche (or soured cream)

$1/4$ cup polenta grains, sieved to remove lumps

$1/4$ cup cornflour

$1/4$ cup finely sliced spring onions

$1/2$ teaspoon each of salt and freshly ground black pepper

Vegetable oil for frying

Steam or boil the corn for 3 minutes, then refresh in cold water. To remove the kernels, hold one end of the cob and rest the other on a chopping board. Run a knife along its length and cut the kernels off, turning the cob as you go. Mix the remaining ingredients (apart from the oil) together well and then stir in the corn. Heat up a frying pan and add a small amount of oil. Drop spoonfuls of the mixture into the pan (but don't crowd it) and cook for 1 minute before flipping the fritters over carefully and frying the other side.

mustard mash

This is a regular side order at the Sugar Club and is unbelievably popular – we've probably served three potato paddocks' worth since we opened! Simple mash is a wonderfully comforting food and adding flavours to it – whether as pesto, olive paste, roast puréed field mushrooms or carrot purée – only increases its appeal. Sue Fletcher at Hodder & Stoughton promised they would publish the *Cookbook* so long as this recipe was included, so here it is. (For photograph see p.126)

FOR SIX SIDE PORTIONS

1kg (2$1/4$lb) boiling potatoes

Salt and freshly cracked black pepper

150ml (5fl oz) double cream

100g (3$1/2$oz) unsalted butter

100g (3$1/2$oz) hot English mustard

Peel the potatoes and cut into quarters. Put them into a deep pot and add 2 teaspoons of salt, cover with cold water and bring to the boil. Simmer to cook, then test with a knife before draining into a colander. Put the cream and butter into the pot and bring to the boil, then add the drained potatoes and mash well. Stir in the mustard and season.

plantain, chilli & polenta fritters

FOR 20 SMALL FRITTERS

225g (8oz) ripe peeled plantain, grated
80g (3oz) polenta grains, sieved to remove lumps
1 egg
60ml (2fl oz) beer
1 teaspoon lemon juice
1/2 cup sliced spring onions
1/4 cup fresh coriander leaves
1/2 teaspoon finely chopped green chilli
1 teaspoon salt
Vegetable oil for frying

Mix all the ingredients together well except for the oil and rest for 15 minutes. Heat up a frying pan and add enough oil to coat the bottom by a few millimetres. Heat to smoking, then carefully add spoonfuls of the mixture and cook for 90 seconds on each side, stacking the fritters on a warm plate when done. As ripe plantain has a high sugar content it can burn easily, so keep an eye on the pan and moderate the heat if necessary.

For those not familiar with plantains, they are those huge 'bananas' that you may have seen in Caribbean or African food stores and markets. They are sold in shades from green (hard, starchy and unripe) to yellow (firm and semi-sweet) to mottled brown (soft and sweet). When ripe they are very banana-like in taste, but their main use is as a source of starch in tropical cooking – and for this the less ripe fruit is normally chosen. If you'd like to try, just boil them in their skins for about 30 minutes in salted water, then peel and mash them with a little oil or butter and some seasoning. It makes a nice change from the usual potato mash.

For this recipe, however, you'll need the very ripe brown plantains. The fritters can be made teaspoon size and used as a canapé – eat them topped with crème fraîche and black olive tapenade. It may seem that combining something from the Caribbean with something Italian is a peculiar thing to do, but it works a treat. Made larger and topped with tomato and coriander salsa they are a delicious starter. You can also make one large fritter and cut wedges to serve as a side dish for grilled pork or roast chicken. A Ghanaian woman, Vasti, first showed me how to turn plantain into a fritter, so thanks to her.

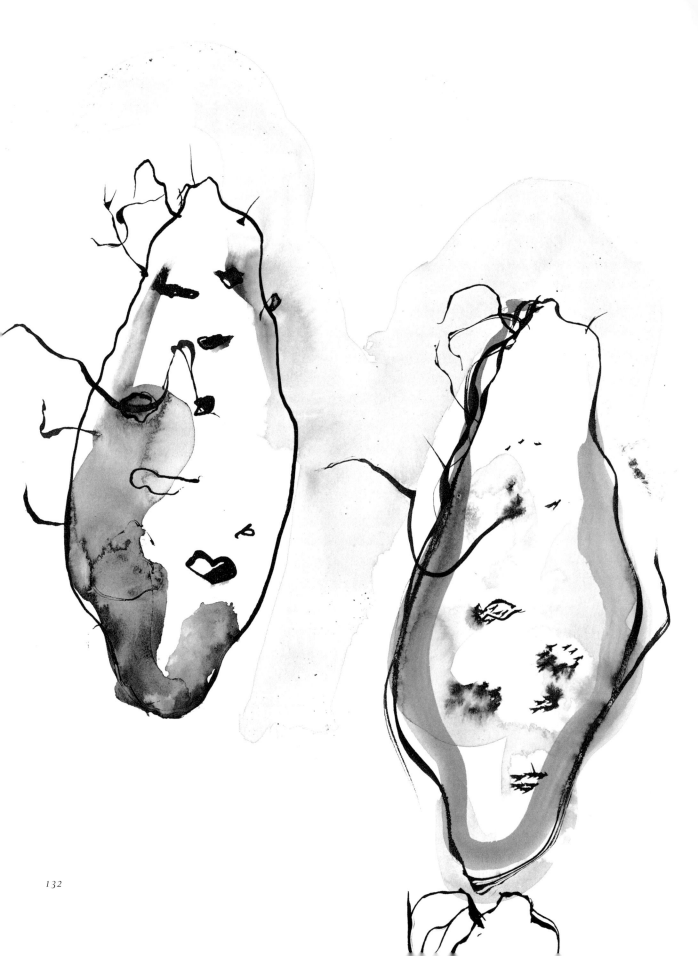

grilled
sweet potatoes

The best sweet potato that we get in England is orange-fleshed and usually comes from Israel or South Africa. The variety we have in New Zealand is called 'kumera' and has a striking purple-coloured flesh and skin. It was the ability to grow and store these tubers that determined how well a Maori tribe survived from one season to the next. In Wanganui we ate them as often as potatoes and cooked in the same way, and in restaurants where I've worked this has been my favourite way of serving them. I grill them on a char-grill, but you can also grill them under a salamander.

FOR SIX SIDE PORTIONS
3 sweet potatoes, each about 350g (12oz)
Salt
Cracked black pepper
Olive oil

Don't peel the potatoes but scrub them gently – the skin keeps the nutrients in. Put them in a deep pot with 2 tablespoons of salt, cover with cold water and bring to the boil. Cooking should take about 20 minutes; when they are almost done a skewer will just go through the centre. Drain the potatoes and leave to cool. Slice into 1cm (1/3in) thick pieces, brush with a little olive oil, sprinkle with salt and pepper and grill for 2 minutes on each side.

braised
fennel
with cumin & ginger

Serve this as a salad on a hot day drizzled with olive and walnut oils, or serve hot in winter with roast cod or leg of pork. For the best flavour choose firm, pale green fennel bulbs. Often all you'll find are the darker green bulbs, but as these sometimes have a ropy texture, make a point of searching out the younger ones that haven't sat around too much. If you're lucky enough to find bulbs with the fronds still attached, keep them on and roast with the fennel in this recipe.

FOR SIX SALAD STARTERS

6 heads of fennel (use only 4 if they're large)
50ml (1³/4 fl oz) lemon juice
¹/2 teaspoon salt
¹/2 teaspoon cracked black pepper
100ml (3¹/2 fl oz) virgin olive oil
2 teaspoons ground cumin
1 medium leek, sliced finely and washed thoroughly to remove grit
2 fingers of fresh ginger, peeled and finely julienned
500ml (18 fl oz) dry white wine

Set the oven to 180°C/350°F/Gas 4. Trim the ends off the fennel and cut each bulb in half lengthways. Then, lengthways again, cut each half into quarters and toss them with the lemon juice, salt and pepper and put in a ceramic roasting dish.

Heat the oil in a 2 litre (3¹/2 pint) saucepan and, when it's hot, add the cumin, fry for 10 seconds, then add the leek and ginger and stir well. Sauté, stirring occasionally, over medium heat until the leek is just beginning to colour, then add the wine. After bringing to the boil, pour the contents of the pot over the fennel and seal with foil. Place in the oven and bake for 40 minutes, then remove the foil and cook for another 20 minutes to brown the fennel.

spicy courgette, spinach & pumpkin
coconut curry

FOR SIX

3 lemon-grass stems

1 large red chilli, stem removed

6 cloves of garlic, peeled

6 lime leaves

3 thumbs of peeled ginger

1 medium-sized red onion, peeled

1/2 cinnamon stick (or 1 teaspoon ground cinnamon)

1 teaspoon freshly ground fennel seeds

1 teaspoon freshly ground cumin seeds

1 teaspoon freshly ground coriander seeds

1/2 teaspoon powdered cloves

100ml (3 1/2 fl oz) sesame oil

600ml (1 pint) unsweetened coconut milk

500g (18oz) fleshy pumpkin or butternut squash,
*   peeled, seeded and cut into 1cm (1/3 in) dice*

4 courgettes, cut into 8mm (1/4 in) slices

Sea salt or Asian fish sauce

600g (1 1/4 lb) washed and picked spinach
*   (if the leaves are large, chop them in half)*

Peel the outer two layers off the lemon grass, discard the top third, remove 8mm (¼in) from the bottom and slice finely. Then put it and all the ingredients down to the cloves into a food processor and purée for 1 minute (if you prefer you can chop them by hand, but this is by far the easiest way).

Heat a 5 litre (9 pint) pot and add the sesame oil. When it's smoking, add the paste and stir well for 2 minutes on a medium to high heat. Add the coconut milk and bring to the boil before adding the pumpkin and simmering for 15 minutes. Add the courgettes and stir well, then cook until the courgettes are almost done (about 10 minutes) – they should still be a little firm. Test for seasoning and add salt or fish sauce to taste. Then stir in the spinach and bring to a gentle boil, remove and serve with plain boiled rice.

This curry base can be used for meat and fish as well – adjust the amount of spices and chilli to suit your taste. The paste can be made up in large quantities and kept, tightly covered, in the fridge for up to two weeks. If you want to freeze it, lay it flat on baking parchment on a baking sheet until solid, then break it up and store in an airtight container. When you need it, take out only as much as you want. Lime leaves, lemon grass and Asian fish sauce are readily available from Thai food stores and some larger supermarkets.

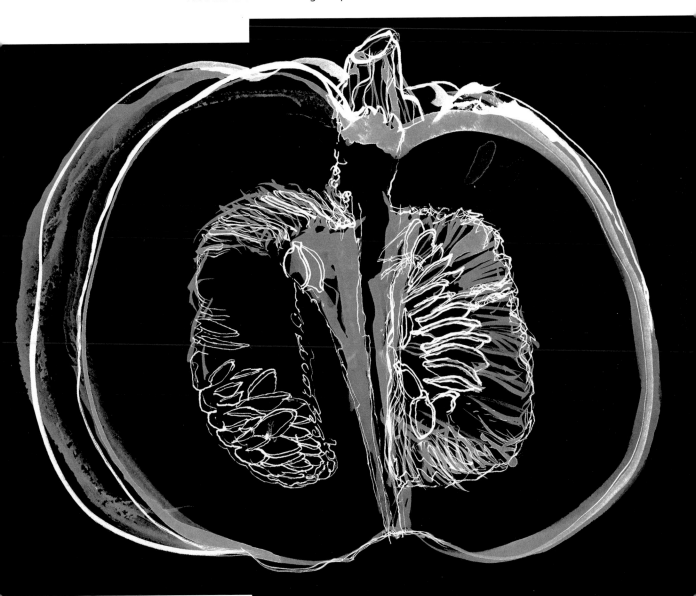

rolled spinach
with pickled ginger,
sesame & tamari

I remember eating this dish at Kuni's Japanese restaurant in Melbourne back in 1983 – although there it was made with dried bonito flakes on top rather than the pickled ginger I use here. Many Japanese restaurants have it on the menu and I tend to eat it whenever I can. It's easy to make and it looks good too. Buy medium-leaf spinach as large leaves can be too thick and the small stuff cooks away to nothing. The amount you need really depends on whether the dish is served as a starter or as a salad to accompany a main meal. A bamboo sushi mat is an essential item for the preparation of this dish.

FOR SIX STARTERS

1–1.5kg (2¼–3¼lb) picked spinach; wash really well and remove the stalks
Salt
80g (3oz) pickled ginger, either home-made (see page 159) or sushi ginger
4 teaspoons white sesame seeds, toasted
4 teaspoons black sesame seeds, toasted
Tamari

Bring a large pot of salted water to the boil and blanch the spinach in batches, cooking only as much as will fit in the pot comfortably. Put a batch in, bring the water back to the boil and then take the spinach out and plunge it into a bowl of cold water, stirring gently. When all the spinach has been cooked and refreshed, tip it into a colander and drain. Take handfuls of it and squeeze as tightly as you can to remove excess moisture. Gently pull these handfuls apart to loosen the spinach.

Now lay your sushi rolling mat, with the bamboo running from left to right, on a clean cloth and put one-third of the spinach on to it. Form the spinach into a long sausage shape that runs along the middle and parallel to the bamboo. Take the end of the mat nearest you and fold it over the spinach, then roll the whole mat up tightly into a tube and roll it back and forwards to shape the spinach. Unroll the mat and do the same to the rest of the spinach.

Cut each roll of spinach into six sections with a very sharp knife and stand the sections on their ends. Place a small mound of pickled ginger on each, sprinkle with the mixed black and white sesame seeds and season with the tamari.

broccoli
& anchovy salad

This makes a wonderful summer salad eaten as a first course or it can be served as an accompaniment to roast chicken or grilled tuna. Salted anchovies are the best to use. Rinse them well under cold water and carefully remove the bones with your fingers. Chop up finely and heat in a few tablespoons of olive oil, squashing gently with a wooden spoon to break them up. If you can't find salted anchovies use the best fillets in oil. Also, if it's available use sprouting broccoli as it both looks and tastes good. Remember that with sprouting broccoli you can eat a lot of the stem as well as the florets.

FOR SIX STARTERS
600g (1¼lb) broccoli
Salt

DRESSING
6 anchovy fillets, preferably salted
1 egg yolk
4 tablespoons seed mustard
60ml (2fl oz) lemon juice
250ml (9fl oz) extra virgin olive oil
½ teaspoon cracked black pepper

The dressing is best made in a mortar with a pestle as the quantity may be too limited for all but the smallest food processors. Put the anchovy fillets (prepared as described above) and egg yolk in the mortar and grind with the mustard to a thick paste. Mix in the lemon juice, then slowly trickle in the oil, mixing well to emulsify it. The consistency to aim for is that of a thick dressing rather than a mayonnaise. Season with the pepper.

Trim and cut the broccoli, keeping the stems long. Steam or blanch in salted water for 2 minutes, then drain and put in a bowl. While still hot, pour the dressing over it and leave for 5 minutes before eating, tossing every minute to 'marinate' it.

brussels sprout
& chestnut purée

FOR SIX GENEROUS SIDE PORTIONS

200g (7oz) dried chestnuts, soaked overnight in cold water
1 bay leaf
600ml (1 pint) milk
1 leek, finely sliced and washed
1 teaspoon fresh rosemary leaves
2 cloves of garlic, peeled and halved
600g (1¹/4lb) brussels sprouts, trimmed of any damaged leaves
and cut into halves
Salt and pepper

Drain the soaking liquid from the chestnuts and throw it away. Put the chestnuts into a deep pot with the bay leaf and pour on the milk, then add hot water to cover them by 4cm (1¹/2in). Bring to the boil and turn down to a rapid simmer with the lid on. After 30 minutes add the leek, rosemary and garlic, adding extra water to cover them if necessary. Replace the lid and continue cooking until the chestnuts are tender; this can take up to 90 minutes in total.

Meanwhile, get a steamer boiling nicely and steam the brussels sprouts until they are cooked. Test by cutting a half sprout in two – it should cut easily. When the chestnuts are cooked drain them, reserving the liquid but discarding the bay leaf. Mix them with the brussels sprouts and purée in a food processor or force through a mouli, adding some of the cooking liquid if the mix is too dry. Season with salt and freshly ground black pepper and serve warm.

I confess to having a problem with this vegetable, maybe due to some childhood memory of badly cooked sprouts. I prefer them wok-fried with lots of chilli and garlic, with each leaf pulled off separately and still crunchy to eat – partly because that way you can't really tell what you're eating! In the recipe here I specify dried chestnuts, but you can use fresh (or tinned as long as they aren't sweetened). The purée is perfect served with a roast or as part of a Christmas feast.

wok-fried
chinese greens
& shiitake mushrooms
with sake

FOR SIX PORTIONS TO ACCOMPANY A MEAL

1.5kg (3¹/4 lb) mixed Chinese greens
10ml (¹/3 fl oz) Asian fish sauce
20ml (³/4 fl oz) tamari
30ml (1 fl oz) sake
30ml (1 fl oz) water
Sesame oil
18 fresh shiitake mushrooms, cut into quarters

Take the greens and discard damaged or discoloured leaves. Remove 1cm (¹/3 in) from the bases and cut the leaves into 10cm (4in) strips, cutting the thicker stems in half lengthways. Wash well and drain. It's important when frying the greens that there is a little water clinging to them, so don't pat them dry. Mix the fish sauce, tamari, sake and water together and keep close by.

Heat up a wok and, when it's smoking, add a few spoonfuls of sesame oil and swirl it around. Add a quarter of the greens and toss gently. Keep on a high heat and fry for 1 minute, tossing occasionally. Add one-fifth of the sake mix and fry for another 30 seconds, then tip the greens into a large bowl. Cook the rest of the greens in the same way.

Finally, heat up the wok again and add sesame oil, then put in the shiitake mushrooms and stir well over a high heat for 30 seconds. Add the last of the sake mix and cook until it is almost evaporated. Add the mushrooms to the greens, mix well and eat at once.

This vegetable dish goes well with so many main courses, such as the pork belly on page 114, barbecued fish, roast chicken or braised duck. For the Chinese greens see what's available in your local Chinatown or Asian supermarket; otherwise use a mix of finely sliced cabbage, chard, celery and courgettes. I tend to use pak choy, bok choy and choy sum, but I will try anything else that's fresh and green. You can use either fresh shiitake mushrooms, which are now easier to find, or the dried variety from Asian supermarkets. Soak dried mushrooms overnight in cold water; next day remove the woody stem and slice the caps finely, discarding the stems. Save the soaking liquid and add some to the wok as you finish cooking the greens. The sake adds a sweet touch to the vegetables, but sweet sherry can be used instead.

deep-fried
baby artichokes
in crispy batter

Baby artichokes (for quantity see note at foot of page)
Plain flour to dust
Vegetable oil for deep-frying
Beer batter (the recipe's on page 83, but halve the amount of paprika)
1 teaspoon fennel seeds, lightly crushed
1/2 teaspoon cumin seeds
1/4 cup finely sliced spring onions

Prepare the artichokes by removing the stems and the half dozen or so lower leaves around the base. Trim sharp-looking bits and cut off the tips if they are a bit woody. Cut them in half lengthways and remove any furry inner parts (though you're unlikely to find these in small artichokes). Toss in flour immediately to prevent discolouring.

Heat up plenty of good-quality cooking oil in a deep pan or deep-fryer to 180°C/350°F. Mix the fennel seeds, cumin seeds and spring onions into the batter and then drop in the flour-dusted artichokes. Fry as many as will fit comfortably in the pan and stir them gently after a minute. Carefully tip over any that are cooking only on one side. They should be ready after 3-4 minutes – they're done when a skewer can be pushed easily through the centre. Drain them on absorbent paper and eat while still warm.

I first made this dish at my eldest sister Vicki's house in Sydney in 1993. We'd found some tiny artichokes on Oxford Street in Darlinghurst, and rushed home to have them for a sunny lunch treat in her mosquito-filled courtyard. These baby artichokes are an excellent starter served with lots of shaved parmesan on top and a bitter leaf salad to the side. They also make a good canapé, especially if you can get them really small. As a side dish to accompany grilled fish they are a treat. I haven't given quantities as how many you need will depend on what you are planning to do with them. For a starter I would serve two small artichokes per portion (by small I mean no bigger than an egg).

pulses

wok-fried
black beans
with pickled red onion & sugar snaps

The colours of this dish are beautiful and the textures are light and crisp. I like to serve it with sliced roast duck breast on top or, as part of a feast, tossed cold through rocket and watercress leaves. The black beans I use are the delicious Spanish type that you find dried in speciality shops rather than the salted Asian variety, although black-eyed beans, aduki beans or mung beans would work as well.

FOR SIX

400g (14oz) dried black beans, soaked overnight in cold water

3 medium-sized red onions

4 large juicy lemons, squeezed to yield 250ml (9fl oz) juice

$^1/_2$ cup chives, cut into 1cm ($^1/_3$in) lengths

1 cup loosely packed fresh mint leaves

$^1/_2$ cup finely sliced spring onions

3 tablespoons sesame oil

1$^1/_2$ cups sugar snaps, topped and tailed

1 teaspoon Asian fish sauce

1 teaspoon tamari

100ml (3$^1/_2$fl oz) water

Rinse the beans, put in a deep pot and add double the volume of cold water; bring to the boil and remove scum. Turn down to a fast simmer and cook for about an hour before testing one; continue cooking if it's still a bit hard. Drain well in a colander when finished. While the beans are cooking peel the onions and slice finely into rings, pour the lemon juice over them, mix well and leave for an hour. Then add to the beans with the juice, herbs and spring onions and mix thoroughly.

Heat a wok to smoking point and put in the sesame oil, swirl around the wok and add the sugar snaps, keeping the heat on high (gas is best for this). Toss the snaps continuously for 1 minute, letting them blacken ever so slightly, then add the beans, toss for another minute, and last add the fish sauce, tamari and water. Cook for another minute and they are done.

red lentil
& parsnip purée

Like mashed potatoes, this purée is one of those dishes that comforts and sustains. Thinned down it makes a great soup, but I like to serve it as a base for oily fish, such as grilled mackerel, sardines or grey mullet. You can vary the taste by adding different herbs, extra garlic or chilli or even by using green lentils in place of red. These will take longer to cook, however.

FOR SIX

2 red onions
150g (5oz) unsalted butter
6 cloves of garlic, peeled and sliced into quarters
500g (18oz) parsnips, peeled and cut into 1cm (1/3in) dice
Salt and pepper
300g (11oz) red lentils, rinsed under cold water
100ml (3 1/2 fl oz) extra virgin olive oil
4 tablespoons fresh oregano leaves
1/2 cup sliced spring onions

Peel and slice the onions thinly and, in a deep pot, sauté in the butter with the garlic for 3 minutes, stirring occasionally. Add the parsnips and 2 teaspoons of salt, cover with cold water and bring to the boil; cook for 10 minutes and add the lentils. Bring back to the boil and add hot water to just cover the lentils, then cook for another 15 minutes, stirring occasionally.

Test a piece of parsnip to see if it's done and, when it is, drain through a colander and reserve the liquid. Put the parsnip mix into a food processor with the olive oil and purée well, adding some of the reserved liquid to bring it to the consistency you want (this will depend on what you intend to use it for). Test for seasoning and gently mix in the oregano and spring onions.

butter bean & herb stew

I love pulses and dried beans. I used to think they were only for winter when fresh beans and peas were unavailable. What a stupid idea! Over the years I've come to depend on them for all sorts of dishes at all times of the year. The following recipe is great whether served hot as a winter lunch with crusty bread and steamed cabbage, served cold on bruschetta at a summer picnic or warmed lightly to serve with barbecued tuna drizzled with basil oil.

As with all dried pulses, don't add salt until the end of the cooking process or the beans will turn out tough – one of cooking's little mysteries.

FOR SIX STARTERS

500g (18oz) dried butter beans, soaked in cold water overnight
2 medium leeks, finely sliced, then washed to remove grit
1 medium-sized white onion, peeled and finely sliced
10cm (4in) stem of fresh rosemary
1 cup loosely packed fresh oregano leaves
12 medium fresh sage leaves
1/4 cup fresh lemon-thyme leaves
8 cloves of garlic, peeled and roughly chopped
1 cup extra virgin olive oil
1/2 cup lemon juice
1 cup roughly chopped fresh parsley
Salt and pepper

Drain the beans and rinse well before putting them in a deep pot and covering with plenty of water. Bring to the boil and cook at a brisk simmer for half an hour, removing foam from the surface. Then add the leeks and onion, the first four herbs, the garlic and the olive oil, stir well and continue cooking for another hour. Make sure there's always at least 3cm (1in) of liquid covering the beans or they may get a little dry. Towards the end test one – the beans should be firm but not crunchy and shouldn't be cooked to a point where they break up. When they're nearly done, add the lemon juice and parsley and cook for a further 5 minutes; test for seasoning and add salt and pepper to taste. The beans are excellent eaten at once but, like most stewed foods, are even better the next day.

black-eyed beans
with sweet potato & coriander

This is a simple dish that's as good eaten cold as a salad as it is warmed up and served as an accompaniment to fish or meat. As with most pulses, you can also thin it down with stock and turn it into a soup.

FOR SIX GENEROUS PORTIONS AS AN ACCOMPANIMENT

2 medium-sized sweet potatoes, peeled
Sea salt
300g (11oz) dried black-eyed beans, soaked in cold water for at least 6 hours
50ml (1³/₄fl oz) olive oil
50ml (1³/₄fl oz) sesame oil
2 red onions, peeled and finely sliced
¹/₂ teaspoon cumin seeds
2 cloves of garlic, peeled and finely chopped
50ml (1³/₄fl oz) lime juice
1 cup fresh coriander leaves

Dice the sweet potatoes into 1cm (¹/₃in) cubes and cover with cold water, then boil until just cooked (be careful not to overcook as they will break up). Drain carefully, keeping the liquid.

Drain the beans and put them into the pot you have just used for the sweet potatoes. Add the potato cooking liquor, adding enough hot water to cover the beans by a few centimetres, and bring to the boil. Skim off any foam that comes to the surface and cook on a gentle boil for at least 45 minutes. Top up with more hot water if it reduces too much. Test the beans to see if they're cooked and continue until they're done, when they should be drained.

Heat up the oils and fry the onion and cumin until the onion softens, then add the garlic and fry for a further 2 minutes. Add the lime juice and bring to the boil, then remove from the heat. Gently mix the sweet potato, beans, onion mix, coriander and 1 teaspoon salt together and leave to sit for at least an hour before using.

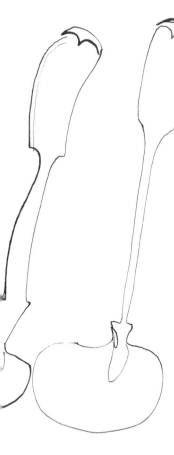

red onion, ginger & basil lentils

FOR SIX GENEROUS SIDE PORTIONS

400g (14oz) lentils
3 medium-sized red onions, peeled and sliced
4 tablespoons sesame oil
200g (7oz) fresh ginger
2 lime leaves
2 stalks of lemon grass
4 cloves of garlic
50ml (1³/4 fl oz) Asian fish sauce
50ml (1³/4 fl oz) tamari
2 loosely packed cups coarsely chopped fresh basil

Wash and rinse the lentils several times in warm water, then drain. Sauté the peeled and sliced onions in the sesame oil over medium heat for 2 minutes, stirring occasionally until they become soft. Add the lentils, cover with cold water and bring to the boil. Meanwhile, peel the ginger and dice it finely; add this to the lentils with the lime leaves and the lemon grass after bashing the grass a bit to release its aroma and oils. Peel and cut the garlic cloves finely and add to the simmering lentils with the seasonings; continue to cook for 30 minutes, making sure the lentils don't dry out at all. Cooking time will depend on the type of lentil you have chosen, so test from time to time. When just soft, remove from the heat and stir in the coarsely chopped basil, check for seasoning and it's ready.

This is an Asian approach to the lentil. Some chefs I've met have argued that the lentil is not an Asian ingredient, but then they probably haven't been to India or gone trekking through Nepal. The lentil is wonderful because it absorbs whatever it's cooked with, whether it be smoked bacon and red wine or ginger and basil. Prepared as in this recipe they're good eaten at any temperature – as a hot side dish with roast chicken, dolloped on to crostini or even mixed into warm rice to serve with a curry. Brown lentils are perfect for this as they retain their shape, as do Puy lentils. Green lentils are fine but they don't look as good.

relishes
& taste adders

All the relishes, chutneys and jams in this section can be stored out of the fridge if you are familiar with preserving and the techniques required to keep food safely for periods of time. With these recipes, though, I have suggested they be stored in the fridge. In this way they will keep for a long time at no risk to you.

jam tomato & chilli

This jam is indispensable in my kitchens both at home and at work. It's great on toast with a fried egg on top, lavished on a slice of roast leg of lamb or pork, or used to glaze a fillet of fish or fowl under the grill. Spread on to bruschetta before adding pink-fried chicken livers, it lifts the whole dish – especially if you sprinkle fresh coriander on top. Or eat it with goat's cheese and rocket in a sandwich.

500g (18oz) very ripe tomatoes, washed
4 red chillies
4 cloves of garlic, peeled
2 thumbs of ginger, peeled and roughly chopped
30ml (1fl oz) Asian fish sauce
300g (11oz) golden caster sugar
100ml (3^1/$_2$fl oz) red wine vinegar

Blend half the tomatoes, the chillies, garlic, ginger and fish sauce to a fine purée in a blender. Although some people have an aversion to seeds, I wouldn't strain this as the tomato seeds provide the pectin that will make the jam set. Put the purée, sugar and vinegar in a deep pot and bring to the boil slowly, stirring all the time. When it reaches the boil, turn to a gentle simmer and add the remaining tomatoes, which you have cut into 5mm (¹⁄₅in) dice, seeds, skin and all. Skim off foam and cook gently for 30–40 minutes, stirring every 5 minutes to release the solids that settle on the bottom. Also, be sure to scrape the sides of the pot during cooking so the entire mass cooks evenly.

When it's done, pour into warmed glass jars and allow to cool to room temperature before storing in the fridge or a cold larder for later use.

pear & date chutney

Chutneys and relishes are such wonderful things to make, to store and also to give away. Like bread, once you've made them a few times you should experiment with your own flavours. If they turn out a disaster, all you've lost is a little time and money, but with luck you'll have learned something that will make your next attempt better.

At Gran's house there would always be different chutneys in the cupboards. Some she had made herself, but others were prepared at Aftercare, an organisation she helped set up to enable mentally disabled people to return to the community. One of Aftercare's great chutney makers was my aunt Mary, who, with Gran's help, also taught me a thing or two.

3kg (6$^{1}/_{2}$lb) ripe pears
1kg (2$^{1}/_{4}$lb) pitted dates
1.5kg (3$^{1}/_{4}$lb) demerara sugar
1 litre (1$^{3}/_{4}$ pints) cider vinegar
200ml (7fl oz) cooking oil
2 tablespoons coriander seeds
2 tablespoons fennel seeds
2 tablespoons cumin seeds
2 teaspoons ground red chilli powder
1.5kg (3$^{1}/_{4}$lb) red onions, peeled and finely sliced
1kg (2$^{1}/_{4}$lb) carrots, peeled and grated
2 tablespoons sea salt

Wash the pears, cut them into quarters and remove the cores, then chop each quarter into three chunks. In a large bowl mix the pears and dates with the sugar and vinegar and leave to sit for an hour.

Heat the oil in a pot large enough to take all of the mixture and, once it's smoking, add the seeds and chilli powder. Fry for half a minute, stirring well, and then add the onions and fry for 3 minutes, stirring continuously. Add the carrots and cook for 5 more minutes. Add the fruit mixture and salt and slowly bring to the boil, stirring constantly. When it comes to the boil turn the heat down to a rapid simmer and cook for 45 minutes, stirring every 10 minutes. Make sure to scrape the bottom and sides of the pot to ensure that nothing sticks or burns.

Heat up some glass or ceramic jars with fairly hot water for a few minutes. Tip out the water and spoon in the chutney almost to the top. Screw on the lids and let the jars cool. Store in the fridge or a cool dark cupboard, where the chutney will keep for a few months.

chilli oil

How simple can a recipe get? This takes care of any extra chillies you have hanging around, but use them only if they're fresh and unblemished. Drizzle the oil into risottos, over fish or meat, or mix it into salad dressings and salsas.

Chillies
Light olive oil

Cut the stems off the chillies and cut them in half lengthways. Put into a pot, cover with oil and place on a high heat. The oil and chillies will soon start to boil and bubble, so stir them a little. Beware, as they tend to pop and will splatter a little oil out of the pot. Once the bubbling slows down – but before the chillies wilt too much – turn the heat off. Allow to cool in the oil before decanting into jars or bottles and leave for a few days before using.

bottled lemons

And here's a way of using up those surplus lemons your friends brought from their organic citrus farm! Watch out for the bubbles that sometimes form when lemons have been in oil for a week or so – this is just the air coming out as they begin to 'pickle'. If this happens you'll need to let the air out every day or two to avoid the fizzing that may occur when you do eventually get round to opening them. Keep the jars in a cool place and use only the best and tastiest lemons. The oil takes on a lovely taste, and after two months you can remove the lemons, dice them and add the dice to soups or stews.

Lemons, washed and wiped dry
Light olive oil

Put the lemons into jars and fill with enough olive oil to cover them completely. Keep them submerged the whole time they are 'pickling' – try placing a small round biscuit cutter on the top lemon to hold it down under the lid. Leave for two months before using.

pickled ginger

Here is a pickle that's really simple to make, but do so only if you can get the freshest
root ginger. If the ginger is stringy it's probably best not to use it for pickling as it won't
be very good. The pickle goes well with raw fish, in sushi or on goat's cheese crostini.

250g (9oz) root ginger
50g (1³/₄oz) caster sugar
400ml (14fl oz) fresh lemon juice

Wash the ginger and scrub it lightly. Slice it as finely as possible, then
sprinkle on the caster sugar and mix well. Add the lemon juice and
mix again. Put into a clean jar and top up with extra lemon juice if
necessary. Seal and leave in the fridge for two days before using

peach,
green chilli & ginger
chutney

1kg (2¹/₄lb) ripe peaches, stones removed and cut into quarters
1 litre (1³/₄pints) cider vinegar
400g (14oz) demerara sugar
250g (9oz) muscovado sugar
500g (18oz) tamarind paste
2 teaspoons ground allspice
3 teaspoons finely ground star anise
500g (18oz) red onions, peeled and quartered
5 hot green chillies, chopped
8 cloves of garlic, peeled and halved
8 lime leaves
4 lemon-grass stems (pull the outer 2 layers off and
  finely chop the lower half of the stems)
300g (11oz) fresh ginger, peeled and finely chopped
2 tablespoons sea salt

In a large pot bring the vinegar, sugars, tamarind and spices to the boil and boil for 5 minutes. Add all the other ingredients and bring back to the boil, then lower the heat to a moderate simmer and cook for 50 minutes until the chutney has reduced to a thick mixture. Be careful to stir gently every 5–10 minutes to prevent sticking on the bottom of the pot.

Spoon the mixture into hot clean jars and seal. Once cool, store in the fridge and leave for at least a week before using. It can be kept for up to six months.

This is a spicy, rich chutney that goes as well with fish as it does with offal and game. It's also good on toast with cheese, or on bruschetta with roast tomatoes and grilled pancetta.

nectarine relish

Not only are nectarines the most mouthwatering fresh fruit, they also make an excellent relish. This relish is great served with cheese, especially a sharp goat's cheese, but it's also good served with cold roast lamb or a game terrine. The quantities given here will make enough to last a long time and you'll find it's easier to prepare in bulk. The recipe is from my sister Donna, the one in our family who preserves the most, and always has a fantastic garden to plunder for produce.

1kg (2¹/₄lb) nectarines, stoned and cut into quarters
500g (18oz) white onions, peeled and finely sliced
1 green chilli, finely chopped
2 teaspoons finely chopped fresh rosemary
100ml (3¹/₂fl oz) cooking oil
200g (7oz) raisins (muscatels go well in this relish)
2 teaspoons crushed coriander seeds
4 teaspoons ground cinnamon
6 cardamon pods, crushed
700g (1lb 9oz) unrefined caster sugar
6 bay leaves
400ml (14fl oz) cider vinegar
100ml (3¹/₂fl oz) Asian fish sauce

Sauté the onions, chilli and rosemary in the oil for 10 minutes over a medium to high heat. Add all the remaining ingredients except the fish sauce and bring slowly to the boil, stirring occasionally to prevent the relish sticking to the bottom of the pan. Turn the heat down and simmer for 1 hour, then add the fish sauce. Continue to cook for about another 30 minutes until the relish is quite thick. Be sure to stir every 10 minutes or so as it thickens or it may catch on the bottom and burn.

When it is ready, spoon into hot, clean jars (rinse them out with hot water beforehand) and seal while still hot. Leave to cool and store in the fridge for at least a week before using. Stored this way the relish will keep for up to six months.

dried fig,
elderflower & lemon relish

600g (1¹/4lb) dried figs, cut into quarters
250ml (9fl oz) cider vinegar
4 juicy lemons
300ml (11fl oz) elderflower cordial
300g (11oz) caster sugar
1 cinnamon stick
2 bay leaves
1 teaspoon paprika
2 teaspoons sea salt

Soak the figs in the vinegar and leave overnight. The next day drain off the vinegar and reserve it. Using a potato peeler, peel the lemons and then juice them. Put the vinegar, lemon peel and juice in a heavy pot and add all the remaining ingredients except the figs and bring to the boil. Boil for 3 minutes before adding the figs and then cook over a moderate heat for 15 minutes, stirring occasionally. Pour into hot, clean jars, seal and cool. Leave for two to three days before using.

This chunky relish goes really well with roast breast of veal or chicken. The elderflower cordial I use is one that you dilute with water to make a refreshing drink. If you can't find any, use good-quality honey as a replacement, but adjust the sugar to suit and add some extra lemon zest.

salsa rossa

You're probably more familiar with salsa verde, the green herb and caper sauce, but this is a really delicious salsa to dollop over grilled fish, mix into a salad with greens or have on hot, oily bruschetta. The recipe makes enough for six large bruschettas.

Mix all the ingredients, adding the seasoning at the end. Leave at room temperature for 30 minutes before using.

3 very ripe tomatoes, cut into 5mm ($1/5$ in) dice
$1/2$ radicchio, washed and cut into 5mm ($1/5$ in) dice
3 red peppers, roasted, peeled and diced
1 small red onion, peeled and finely diced
$1/2$ cup shredded fresh basil leaves
$1/4$ cup shredded fresh mint leaves
$1/4$ cup finely sliced chives
$1/4$ cup chopped fresh chervil
350ml (12fl oz) extra virgin olive oil
50ml ($13/4$fl oz) lemon juice
Salt and freshly ground black pepper

salsa verde

1 cup fresh flat-leaf parsley leaves
$1/2$ cup fresh chervil leaves
1 cup fresh basil leaves
1 cup fresh mint leaves
1 tablespoon fresh thyme leaves
$1/4$ cup small capers, rinsed
$1/4$ cup cornichons, rinsed
$1/2$ cup finely sliced chives
$1/2$ cup finely sliced spring onions
350ml (12fl oz) extra virgin olive oil
Salt and freshly ground black pepper
50ml ($13/4$fl oz) lemon juice

Chop the first five herbs along with the capers and the cornichons (by hand or in a food processor). Gently mix in the remaining ingredients, adding the juice last of all to prevent the mixture discolouring. Leave for half an hour to let the flavours develop before using.

This sauce is traditionally supposed to be served only with poached meats and fish – a rule with which I strongly disagree as it goes counter to my own much better rule that if something tastes right to you it's likely that others will find it good too! It goes particularly well with barbecued meats, fish and vegetables, whose smoky flavours are strong enough to compete with the tastes of the salsa. The recipe will make a large jarful.

smoked paprika, feta & almond pesto

This is one of those Sugar Club items that owes its origin to a mistake. The usual basil and pine-nut pesto (served on grilled squid) was on the menu, but it had run out. As I was preparing to make a new batch panic set in – we had no pine nuts, basil or parmesan. On the spot I had to come up with something that would go with the dish and could be made from ingredients that we did have to hand. This is the result, and it's now a regular on the restaurant's ever-changing menus. Smoked paprika is a staple flavouring in Spain and is what gives most chorizo its taste. You should be able to find it in any good food hall or Spanish delicatessen.

Purée the first four ingredients to a coarse paste in a food processor. Crumble the cheese and add with half the oil and purée for 30 seconds. Scrape down the sides of the bowl, add the remaining oil and the lemon juice and purée for a further 15 seconds. Tip into a bowl and stir in salt to taste. If you want a thinner pesto, just mix in more oil.

2 tablespoons smoked sweet paprika
4 cloves of garlic, peeled
300g (11oz) shelled almonds, toasted
1 cup roughly chopped parsley
200g (7oz) feta cheese
700ml (24fl oz) olive oil
50ml (1³/4fl oz) lemon juice

brown roast chicken stock

MAKES ABOUT 4 LITRES/7 PINTS

1 boiling chicken, cut into 8 pieces
2 onions, cut with the skin into quarters
2 carrots, peeled but not cut
¹/2 head of celery, cut in half
1 head of garlic, unpeeled, cut in half horizontally
1 10cm (4in) thumb of fresh ginger, cut into 6

6 bay leaves
¹/2 cup fresh herbs (e.g. rosemary, sage, oregano, tarragon – you choose!)
100ml (3¹/2fl oz) cider vinegar
750ml (26fl oz) red wine
4 litres (7 pints) water

Preheat the oven to 220°C/425°F/Gas 7. Put the chopped-up chicken and the onions in an oven dish and roast for 1 hour, stirring from time to time. After 40 minutes add the carrots, celery and garlic. Once the bones have gone a dark brown colour – but not burnt – tip them and the vegetables into a large pot (at least 8 litres or 14 pints). Add the rest of the ingredients and bring to the boil. Pour an extra 600ml (1 pint) boiling water into the roasting dish and leave to soak for 30 minutes, then scrape the dish and add the contents (water and all) to the pot as well. Skim any scum from the surface of the stock and simmer for 3-5 hours. Strain well and it's ready to use.

pesto

A 'pesto' is simply a sauce, and there are as many ways to make it as there are cooks. In some parts of Italy – especially Liguria, where it originally comes from – it is only made with pecorino sardo, a delicious sharp ewe's milk cheese. Parmesan wouldn't even get a look in. Personally, what I'm serving it with influences how much of each ingredient I put into it. I love pesto with grilled fish, when I use a little less parmesan. For roast pumpkin I add more parmesan. I also love it thinned down with olive oil as a salad dressing, when I reduce the amount of garlic and add a little lemon juice. And so on. The following recipe will make enough for a good-sized jar of tasty pesto.

10 cloves of garlic, peeled
3 loosely packed cups fresh basil leaves
1 loosely packed cup fresh mint leaves
1 cup roughly chopped fresh flat-leaf parsley leaves
500ml (18fl oz) olive oil
1 cup pine nuts, lightly toasted to a golden brown and left to go cold
1 cup finely grated parmesan
Some more olive oil

Put the garlic, herbs, olive oil and pine nuts into a food processor and purée to a coarse paste, but be careful not to overwork it. Turn into a mixing bowl and stir in the parmesan. Stir in more olive oil to get the consistency you want.

labne

This is simple to make and it comes in handy for all sorts of things – baking with brains, mixing with chillies and lime zest or dolloping on a spicy piece of grilled chicken. The name comes from the Arabic for white.

400ml (14fl oz) thick natural yoghurt

Line a colander or a sieve with a piece of clean calico or a new J-cloth. Tip the yoghurt in and fold the cloth over the top. Put the sieve or colander into a bowl and place in the fridge to drain. After 24 hours tip the contents of the cloth into a clean bowl. Your labne is now ready to use, but do keep it stored in the fridge. The drained liquid can be used to make bread with.

desserts

fruit & nut biscotti

This recipe is from Katherine Smyth, a fellow New Zealander, who came across it while cooking in Sydney, Australia. She is a great friend, a terrific chef and a brilliant potter. Many of the plates and bowls in the illustrations were made by her especially for the book. I first met Kath in New Zealand when the original Sugar Club opened, where she came to cook with us. Now based in London, she still cooks with us a little but devotes most of her time to potting.

These biscuits have become a regular item on the Sugar Club menu and are usually served with mascarpone and fresh fruit. Mangos and berries are particular favourites, and the caramelised peaches on page 173 are great too. The biscotti are very hard, so try them dipped in Vin Santo, coffee or tea. As the recipe makes a big batch, store the surplus in airtight jars and they'll keep for three months. At the Sugar Club we make them with assorted dried fruit and nuts, but they can be made with just almonds or a nut of your choice. (For photograph see page 166.)

500g (18oz) plain flour
500g (18oz) unrefined caster sugar
1 tablespoon baking powder
5 eggs, lightly beaten
100g (3¹/₂oz) plump sultanas
100g (3¹/₂oz) dried apricots, sliced

100g (3¹/₂oz) pitted dates, chopped
100g (3¹/₂oz) shelled pistachio nuts
100g (3¹/₂oz) whole blanched almonds
100g (3¹/₂oz) shelled hazelnuts
2 lemons (grate the zest and discard the fruit)

Preheat the oven to 180°C/350°F/Gas 4. Mix the flour, sugar and baking powder in a large bowl. Add half the beaten eggs and mix well, then add half of what's left and mix again. Now add the last quarter a little bit at a time until the dough takes shape but isn't too wet (you may not need to use all of the eggs). Add the fruit, nuts and lemon zest and mix well.

Divide the dough into six, roll into sausage shapes about 3cm (1in) in diameter and place, at least 6cm (2¹/₂in) apart, on baking parchment on baking trays. Wetting your hands when rolling these out helps to prevent the dough sticking. Lightly flatten the 'sausages' and bake until golden brown (20-30 minutes). Remove from the oven and leave for 10 minutes to cool and firm up.

Drop the temperature of the oven to 140°C/275°F/Gas 1. With a serrated knife, cut the biscotti on an angle into 5mm (¹/₅in) slices and lay these on the baking trays. Return them to the oven and cook for 12 minutes, then turn the biscotti over and cook until they are a pale golden colour (10-15 minutes). When ready, remove from the oven and cool on cake racks. Store in airtight jars.

pistachio
sbricciolona

This biscuit is a version of the original from Ferrara in Italy. Sbricciolona means 'she that crumbles' and is an apt description. At the restaurant we serve the biscuits with something creamy and something fruity. One day it may be roast apricots and mascarpone, another it may be raspberries and whipped cream. You can also substitute another nut for the pistachios – in my earliest version it was almond.

130g (4¹/₂oz) shelled pistachio nuts
170g (6oz) plain flour
120g (4oz) light brown soft sugar
100g (3¹/₂oz) polenta grains, sieved to remove lumps
Finely grated zest of 1 orange
2 egg yolks
130g (4¹/₂oz) unsalted butter, melted but cooled

Turn the oven to 180°C/350°F/Gas 4. Grind the nuts to fine crumbs in a food processor. Add the next four ingredients and process for 10 seconds, then add the egg yolks and process for another 10 seconds. Last, add the cooled butter and process for 20 seconds. Tip on to a baking parchment-covered baking tray and spread loosely to a depth of 8mm (¼in). Place in the oven and bake to a golden colour (about 40 minutes, checking after 30 minutes).

Remove from the oven and leave to cool for 5 minutes before cutting the sbricciolona into 1cm (¹/₃in) wide fingers. Allow to cool completely before transferring from the tray to a cake rack. After 1 hour store in an airtight tin.

chocolate & star anise
mousse cake

This is more like a chocolate mousse that has been baked to form a crust. It's flourless, which is perfect for people on a gluten-free diet, and it's also incredibly rich. You can vary the flavours – adding 1 cup of ground roasted hazelnuts to the egg yolks in place of the star anise works well. Serve with plain whipped cream or crème fraîche, or it goes down a treat with tamarind ice cream, a combination that brings east and west together deliciously. Here I make the choice for you!

MOUSSE CAKE

*300g (11oz) dark bitter chocolate
(minimum 60% cocoa fat)*
150g (5oz) unsalted butter
6 eggs
1 1/2 teaspoons freshly ground star anise, sieved
50g (1 3/4oz) caster sugar

TAMARIND ICE CREAM

7 egg yolks
200ml (7fl oz) honey
300ml (11fl oz) milk
300ml (11fl oz) double cream
200ml (7fl oz) tamarind paste

Line the base and sides of a 20cm (8in) spring-bottomed cake tin with greaseproof paper and set the oven to 180°C/350°F/Gas 4. Melt the chocolate and butter in a metal bowl over a pan of simmering water. Separate the eggs and whisk the yolks with the star anise and 2 tablespoons of the sugar for half a minute. Stir in the melted chocolate and mix well. Beat the egg whites with the remaining sugar until very stiff, quickly fold one-third of the whites into the chocolate mix, then gently fold in the remainder and pour the mix into the cake tin. Place on the middle shelf of the oven and bake for 20 minutes. Remove from the oven and cover the tin with foil, sealing well to keep the heat in so that the steam can soften the crust. Once it's cold put in the fridge and leave for at least 4 hours before giving in to temptation.

To make the ice cream, whisk the yolks and half the honey together for 1 minute. Bring the milk, the cream and the remaining honey to the boil in a large pan and gently whisk into the egg yolks. Return the mix to the pan and stir continuously on a low heat for 4-6 minutes until the custard coats the back of the spoon. Pour into a bowl and mix in the tamarind paste. Stir every few minutes until cool, then leave in the fridge to go completely cold. Churn in an ice-cream machine to the manufacturer's instructions.

chocolate pots

The chocolate pots are easy to make and are best prepared a day or two ahead. They shouldn't be made too large as they are very rich. The richness will depend on the quality of the chocolate – the more bitter it is, the better the result will be, and I recommend going for chocolate with a cocoa butter content of between 55 and 70%. The recipe is adapted from one that was used at the old royal courts of France.

FOR SIX MEDIUM-SIZED RAMEKINS

400ml (14fl oz) milk
250ml (9fl oz) double cream
200g (7oz) bitter chocolate, coarsely grated
4 egg yolks
45g (1 1/2 oz) caster sugar
1 teaspoon vanilla essence
Lightly whipped cream for serving

You'll need six 250ml (9fl oz) ceramic ramekins and a roasting tray as deep as the ramekins and just big enough to hold them. Bring a kettle full of water to the boil and turn the oven to 170°C/340°F/Gas 3–4.

Heat the milk and cream and, when it comes to the boil, add the chocolate, stirring gently until it melts. In a large bowl whisk the yolks, sugar and vanilla. Gently whisk the chocolate mixture into this and transfer to a heatproof jug. Arrange the ramekins in the roasting tray and divide the mixture between them. Place in the oven and gently pour in the hot water until it comes two-thirds of the way up the sides of the ramekins. Cook for 30 minutes before testing.

When the chocolate pots are ready they will still be runny in the middle but slightly crusted on top. Remove from the oven and leave to cool in the water. Cover the ramekins with clingfilm and keep in the fridge until needed.

Serve the chocolate pots with a great dollop of cream on top and a cumin shortbread (see opposite) or an almond wafer (see page 177).

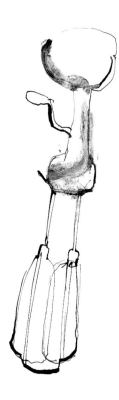

cumin shortbread

This recipe, apart from the cumin, is my grandmother Molly's. I find the cumin goes so well with the richness of the chocolate pots. She used salted butter, I use unsalted; use whichever you prefer. The amounts given below will make a lot, so keep any extra for afternoon teas!

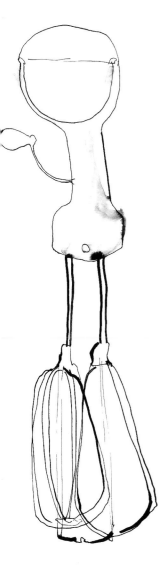

250g (9oz) butter, at room temperature
3/4 cup caster sugar
1 cup cornflour
2 cups plain flour
2 teaspoons freshly ground cumin powder
3/8 cup finely ground almonds

Cream the butter and sugar and sift the flours and cumin. Mix half the flour and half the almonds into the butter and sugar, then add the rest and knead gently until the mix holds together in a firm ball. Roll out with a rolling pin and cut with a biscuit cutter (or form into balls and press out with a fork) and place on a greased baking tray. Bake for 20 minutes in an oven that's been preheated to 160°C/325°F/Gas 3, then check. The biscuits should be golden, not brown, and just cooked in the middle. Remove from the oven and allow to cool for a few minutes, then transfer to a cake rack to cool completely before storing in an air-tight container.

caramelised peaches

6 ripe peaches, cut in half and stoned
1/4 cup water
2 cups caster sugar

Put the water in a deep frying pan or saucepan (large enough to hold all the fruit), add the sugar and stir over a moderate heat until it has dissolved. Turn the heat up to high – but don't stir from now on – and boil until the syrup begins to turn a golden colour. Then carefully add the peach halves and a few tablespoons of hot water, taking care as this can spit a little. Turn the heat down to moderate. Cook the peaches first with the cut side up for 4 minutes, then turn them over and cook until done (they should still feel a little firm when you insert a skewer). Let them cool in the liquid, then remove the skins.

The peaches are excellent with the pashka on page 175.

I love to eat pashka on cold nights with stewed fruit or a simple biscuit, or both. The caramelised peaches on page 173 are a wonderful accompaniment. The pashka can be made in one large mould or in several smaller ones. Small terracotta plant pots with a hole in the bottom look good and, as you will find, the hole is needed to allow liquid to drain from the dessert. The biscuits are based on a type I used to eat a lot of as a child. The original recipe is from the *Edmonds Cookbook* – that essential reference in all New Zealand kitchens.

pashka with
ginger melting moments

FOR SIX GENEROUS PORTIONS

PASHKA

200ml (7fl oz) milk

60g (2oz) unrefined caster sugar

4 egg yolks

1/2 teaspoon vanilla essence

50g (1 3/4oz) plump raisins

50g (1 3/4oz) currants

250g (9oz) cottage cheese

275g (10oz) cream cheese

80g (3oz) unsalted butter, at room temperature

50g (1 3/4oz) shelled pistachio nuts

GINGER MELTING MOMENTS (MAKES 16)

200g (7oz) unsalted butter,
    at room temperature

60g (2oz) icing sugar

125g (4 1/2oz) plain flour

125g (4 1/2oz) cornflour

1/2 teaspoon baking powder

2 tablespoons peeled and finely
    grated or chopped fresh ginger

Make the pashka first. Heat the milk and half the sugar in a pan, bring to the boil and remove from the heat. Whisk the remaining sugar, the egg yolks and the vanilla together and whisk the milk into this, mixing well. Return the mixture to the pan and stir continuously with a spoon over a moderate heat until the mix is thick enough to coat the spoon. Be careful not to overcook or it will curdle. When the consistency is right, pour it into a large bowl, add the raisins and currants and leave to cool to room temperature.

In a mixing bowl, beat together the cottage cheese, cream cheese and butter for a minute, then stir in the cooled custard and the nuts. Line a colander or individual moulds (with drainage holes) with calico or J-cloth and fill with the mixture. Fold the cloth over the top, place the mould on a plate and leave in the fridge for at least 12 hours to drain and set.

For the biscuits, preheat the oven to 170°C/340°F/Gas 3–4. Cream the butter and sugar, then sift the flours and baking powder and add to the butter along with the ginger. Mix gently until the dough forms a ball. Divide this into quarters, then quarter again to give 16 pieces. Roll each into a ball and place, 4cm (1 1/2in) apart, on baking parchment on a baking tray. Press the balls down with your fingers and put in the oven for 15 minutes, but don't let them colour too much. When done, remove from the oven and allow to cool on the tray for a few minutes, then transfer to a cake rack to cool fully. Store in airtight jars.

ann o'carroll's
treacle tart

Oh, if I had a penny for every time a customer has said this is the best treacle tart they've ever had! Annie who is from Wales, is the senior sous-chef at the Sugar Club. She's been with us almost from the start and is a bit of a character – a very talented one too. I asked her for a recipe for this book, and this is what she came up with. My only input is to suggest making the pastry with lemon zest as I feel it helps cut through the richness of the filling. So here's how to make the most delicious treacle tart to be found in London!

The recipe will make one large tart for eight people. Use a 28–30cm (11–12in) loose-bottomed tart tin. There may be more mixture than you need, so don't fill the tart shell too full as it will overflow and stick the tart to the tin. It can then be almost impossible to remove without ruining the crust. Golden syrup and treacle are one and the same, by the way – hence the absence of 'treacle' in the ingredients list.

SWEET-LEMON PASTRY

300g (11oz) unsalted butter, very cold and cut into 1–2cm (1/2in) cubes
500g (18oz) plain flour
2 teaspoons finely grated lemon zest
200g (7oz) icing sugar
1 egg, lightly beaten
50ml (1³/4fl oz) cold milk

FILLING

4 eggs
Finely grated zest and juice of 1 lemon
550ml (19fl oz) golden syrup
450ml (15fl oz) double cream
150g (5oz) brioche crumbs (or bread or croissant crumbs)
1 medium apple, unpeeled, cored and grated

Pulse the butter and flour in a food processor for 15 seconds, then add the lemon zest and icing sugar and pulse for 10 seconds. Add the egg and half the milk and pulse until it forms a ball. It is important not to overwork the pastry as it will toughen – if it's too dry add a little more milk. Remove the pastry and knead it gently for a few seconds, then wrap it in clingfilm and leave in the fridge for 30 minutes.

Take the base from the tart tin and press the pastry on to this, spreading it with the back of your hand. Dust the worktop and the pastry with flour and roll the pastry out into a circle until it is 3–5mm (1/10-1/5in) thick; the circle should be large enough to cover the sides of the tin. Carefully lift the pastry and base from the worktop and sit in the tart tin. Press the pastry gently into the tin, trim off any excess and put the case in the fridge or freezer for at least an hour. Then line the pastry with baking parchment, fill with cooking beans and bake for 15 minutes in an oven that's been preheated to 180°C/350°F/Gas 4. Remove the beans and paper and cook for a further 10-15 minutes until the case is an even golden colour.

For the filling, set the oven at 160°C/320°F/Gas 3. Whisk the eggs, lemon juice and zest for half a minute, then add the golden syrup and cream and whisk until the mix emulsifies. Finally, add the crumbs and the apple and mix well. Pour the mixture immediately into the pre-cooked tart case, place on a baking tray and put in the oven. Check after 35 minutes. The tart is cooked when the filling has gone golden and slightly puffy. Rock the tart on its tray: it should wobble but show no sign of uncooked filling. If it does, return to the oven until done.

almond wafers

These crunchy wafers can be made with almost any nut, including coconut. The rule to follow is that if you use finely ground nuts, use half the quantity that you would if they were flaked. Any spice can be added to the mix (e.g. ground cinnamon or nutmeg), so it's a versatile recipe and a great way to use up egg whites. The mix keeps in the fridge for up to ten days.

FOR ABOUT 15 WAFERS

300g (11oz) almond slivers (or half that weight of ground almonds)
270g (9^1/$_2$oz) caster sugar
60g (2oz) plain flour
240g (8^1/$_2$oz) egg whites
100g (3^1/$_2$oz) unsalted butter, melted and still slightly warm

Mix the dry ingredients well in a large bowl. Thoroughly stir in the egg whites, then thoroughly mix in the butter and leave for 20 minutes at room temperature.

Preheat the oven to 180°C/350°F/Gas 4. Cover a baking tray with non-stick baking parchment (or grease it well) and dollop 4–6 dessert-spoonfuls of the mix on to the tray, though not too close together as they spread. Using the back of a fork that's been dipped in cold water, spread each lump into a disc about 10cm (4in) across. Cook in the oven until an even golden colour (about 12 minutes). Remove and, with a metal spatula, take the wafers from the tray and allow to cool on a cake rack. You can also drape them over a rolling pin to form what the French call *tuiles* ('roofing tiles'). Store the wafers in an airtight jar.

coconut tart
with
passionfruit mascarpone

Despite its simplicity this tart is rich and delicious. It's also dead easy to make. The mascarpone and passionfruit are a combination that finish it off just so, but a bowl of whipped honey–yoghurt cream will also do the trick when passionfruit aren't around.

1 25cm (10in) sweetened shortcrust pastry tart shell, blind-baked
2 eggs
2 lemons, the zest grated and the fruit then juiced
200g (7oz) caster sugar
375ml (13fl oz) double cream
3 cups desiccated coconut

PASSIONFRUIT MASCARPONE
250g (9oz) mascarpone
250ml (9fl oz) double cream
100g (3¹/₂oz) golden caster sugar
150ml (5fl oz) passionfruit pulp

Preheat the oven to 160°C/325°F/Gas 3. Mix the eggs, lemon zest and caster sugar for one minute. Gently mix in the cream, then the lemon juice and finally the desiccated coconut. Pour into the pastry shell and bake until golden all over (about 40 minutes). Remove and cool – leave for 1 hour to firm up.

To make the accompaniment, let the mascarpone warm to room temperature before putting it into a bowl with the double cream and sugar. Whip until it begins to thicken, add the passionfruit pulp and beat until firm.

pear & amaretti tart

FOR EIGHT

1 30cm (12in) shortcrust pastry tart shell, blind-baked
210g (7¹/₂oz) unsalted butter
90g (3oz) unrefined caster sugar
250g (9oz) amaretti biscuits, crushed
110g (4oz) ground almonds
4 free-range eggs
Finely grated zest of 2 lemons
1 tablespoon lemon juice
6 firm sweet pears

There are two ways of making this tart. The easiest uses a food processor and doesn't require the butter to be soft. Put the butter, sugar, amaretti and almonds into the processer and purée to a paste for 45 seconds. Add the eggs, zest and juice and process for another 30 seconds, processing again briefly after scraping down the bowl.

The second method is by hand. Have the butter at room temperature and cream it with the sugar, then whip in the amaretti and almonds. Beat the eggs, zest and juice lightly and whisk this gradually into the butter mixture until it is all absorbed.

Set the oven to 180°C/350°F/Gas 4. Spread the mix into the base of the tart shell, levelling it out to come halfway up the case. Cut the pears (peeled if you prefer) into quarters and remove the cores, then place them in a circle evenly around the base, pressing them into the mixture. Cook for 40 minutes in the middle of the oven. If the tart starts to go too brown, cover lightly with a piece of foil. Allow to cool in the tin before removing.

Amaretti are those delicious little crunchy biscuits from Italy that are often thought to be made with almonds. The unusual taste actually comes from ground apricot kernels – from which the well-known liqueur is also made. This pear and amaretti tart is rich and decadent and should be served with lots of lightly whipped cream. It is fine prepared a day ahead.

sticky toffee pudding
with caramel sauce

What is so Pacific about this pudding, you may well ask. Absolutely nothing at all! I include it here because it's such a delicious and light dried-fruit pudding, and it's a lovely warmer. If you like, add a mashed ripe banana and an extra 2 tablespoons of flour when mixing in the flour and pretend you're in Samoa.

FOR SIX

160g (5½oz) demerara sugar
130g (4½oz) unsalted butter,
* at room temperature*
1 teaspoon vanilla essence
1 egg
250g (9oz) plain flour
1 teaspoon baking powder
300ml (11fl oz) water
200g (7oz) pitted whole dried dates
50g (1¾oz) currants
50g (1¾oz) sultanas
100g (3½oz) walnut pieces
1 tablespoon bicarbonate of soda

CARAMEL SAUCE
250g (9oz) caster sugar
100ml (3½fl oz) water
100ml (3½fl oz) orange juice
300ml (11fl oz) double cream

For the puddings, set the oven to 180°C/350°F/Gas 4 and lightly oil six 300ml (11fl oz) ramekin dishes. Cream the sugar and butter until pale-coloured, then add the vanilla and egg and beat again for a minute. Sift the flour and baking powder, mix in, and set the batter aside in a warm place. Put the water, fruit and nuts in a pan and bring to the boil, then remove from the heat and stir in the bicarbonate of soda (don't worry about the frothing). Stir into the batter and mix well. Spoon the mixture into the ramekins until three-quarters full, place on a baking tray and put in the oven. Test after 25 minutes by inserting a skewer into the puddings; it should come out clean, though if a little fruit sticks to it that's fine. If the batter is still undercooked, return to the oven until done. Once cooked, let the puddings sit in their dishes for 10 minutes before turning out.

Now for the caramel. Put the sugar and water in a deep pot and bring to the boil, stirring continuously. Once it comes to the boil do not stir; keep on a high heat and soon it will start to go golden, then brown. As soon as there's the least whiff of burning remove the pot from the heat and pour in the orange juice. It will bubble and steam frantically, so stand clear. When it settles down, return the pot to the heat and return to the boil. After 1 minute add the cream and bring to the boil again. Pour the caramel sauce over the toffee puddings and serve with lashings of lightly whipped cream.

vattalapam

This is a dessert I first made at Rogalsky's restaurant in Melbourne. I'd been reading through some exotic cookbooks and came across a dish that was described as the 'national dessert of Ceylon'. Over the years I've asked numerous Sri Lankans if this is something they have at all often. Some say yes, others no. Anyway, it's a fantastically rich and densely flavoured dessert and serving it with mango or papaya helps to cut through that richness – though my advice is to pile on the luxury by including a generous dollop of cream!

FOR SIX

500ml (18 fl oz) unsweetened coconut milk
3 teaspoons finely ground green cardamon
1 teaspoon finely ground mixed spice
100g (3¹/₂oz) golden syrup (coconut-palm sugar in the original recipe –
try an Asian food store and use if you can find it)
5 eggs (please always buy free-range)
¹/₂ teaspoon pure vanilla essence
80g (2³/₄oz) demerara sugar (again, use coconut-palm sugar if possible)
2 tablespoons plain cooking oil
60 unsalted cashew nuts

Preheat the oven to 180°C/350°F/Gas 4. Have six 200ml (7fl oz) ovenproof ramekins ready (ceramic soufflé dishes are ideal). Prepare a *bain-marie* by half filling a roasting dish with hot water and place it on a sheet in the top third of the oven. Bring the milk, spices and golden syrup to the boil, cover with a lid and put aside in a warm place for 10 minutes. Beat the eggs, vanilla and demerara sugar for 30 seconds and gently whisk the hot milk mixture into them. Oil the ramekins and divide the cashew nuts between them, then pour the custard over the nuts. Place in the *bain-marie* and bake for 40 minutes.

Test the vattalapam by inserting a wooden skewer; it should come out clean, if a little moist. Let them cool in the ramekins before placing in the fridge for at least 2 hours to set (they will keep for up to four days). Run a knife around the sides and shake gently to remove the dessert. Serve with wedges of fresh tropical fruits.

upside-down
pear & polenta cake

CAKE

380g (13¹/₂oz) demerara sugar
   (and some for sprinkling)
4 eggs
1 vanilla pod (seeds only)

250ml (9fl oz) vegetable oil
250ml (9fl oz) chardonnay
300g (11oz) plain flour
2¹/₂ teaspoons baking powder
120g (4¹/₄oz) polenta grains,
   sieved to remove lumps
6 sweet pears

TOPPING

2 cups caster sugar
¹/₂ cup water
3 pears, peeled, cored and
   diced into 1cm (¹/₃in) pieces
Juice and zest of 2 lemons

Set the oven to 180°C/350°F/Gas 4. Line the base of a deep 25cm (10in) spring-form cake tin with baking parchment and sprinkle with 3 tablespoons of demerara sugar. Beat the eggs, vanilla seeds and sugar for 1 minute, then beat in the oil and the wine for 30 seconds. Sift the flour, baking powder and polenta, add and mix well. Peel the pears, cut into sixths and core. Arrange the pears on the sugar in the cake tin in a fan pattern, pour on the batter and put in the oven. After 30 minutes cover the tin loosely with foil and cook for another 30 minutes. Test by inserting a skewer – keep cooking until it comes out clean. Let the cake cool in the tin for 30 minutes, then turn out and pour on the topping.

   To make the topping, bring the sugar and water to the boil in a deep-sided pot, stirring to dissolve the sugar. Once it's dissolved keep on a vigorous boil but don't stir. Cook until the syrup turns into a dark golden caramel, then gently but quickly add the pear pieces. Cook for another 2 minutes, stirring the mixture gently, before adding the lemon zest and juice. Turn to a simmer and cook until the pears are tender.

This cake is a particular favourite of mine and I have made it in every restaurant I've worked in in London. I invented it because of a desire to have more dairy-free desserts in my repertoire when I was on a dairy-free diet. The irony is that it's best served with lashings of whipped cream! I've also made a successful version using rhubarb in place of the pears. Adding ground spices to the batter is another way of varying the taste.

boiled orange
& brazil nut cake
(with or without chocolate ganache)

This cake is easy to make and keeps moist for a long time. Also, it's wheat- and dairy-free, which means we are able to surprise dietarily-restricted diners at the restaurant with that rare treat for them – a dessert. I first made it in 1983, in Melbourne, when the recipe was given to me by a great chef friend, Brie (real name Mareesa Mayne). She was working with an Egyptian chef who gave her his mother's favourite cake recipe. So it's with many thanks that I pass on to you my version of her version of his version of a Middle Eastern mother's orange cake.

CAKE

3 medium-size oranges
1 lemon
250g (9oz) caster sugar
300g (11oz) finely ground, lightly toasted Brazil nuts
6 eggs
2 teaspoons baking powder
1 teaspoon ground cinnamon
1/2 teaspoon ground allspice

CHOCOLATE GANACHE

1 1/2 cups grated bitter chocolate
1 1/2 cups double cream, straight from the fridge

Wipe the fruit well and place in a pot with enough cold water to cover. Bring to the boil and simmer briskly with the lid on for 20 minutes, topping up with hot water if the fruit isn't floating. Remove from the pan and cool for 20 minutes.

Preheat the oven to 170°C/340°F/Gas 3-4. Cut the boiled fruit in half and remove seeds, then place in a food processor, skin and all. Add the remaining ingredients and purée for 30 seconds; wipe the sides of the bowl back into the mix and purée for another 20 seconds. Pour the thick batter into a spring-form cake tin and bake for about 50 minutes. Check after 40 minutes by inserting a thin knife into the cake. It may come out with a tiny bit of batter, but so long as it is cooked batter you are all right – remember, this is a moist cake! Remove from the oven and, when cool, take from the cake tin and leave to sit for a few hours before eating.

This cake is delicious enrobed with chocolate ganache. Gently melt the grated bitter chocolate in a heatproof bowl over a pot of boiling water. Stir in the double cream and spread on the cold cake.

poached tamarillo
with vanilla-yoghurt bavarois

FOR SIX DESSERTS

6 tamarillos

750ml red wine (standard bottle size)

200g (7oz) demerara sugar

2 star anise

1 cinnamon stick

BAVAROIS

150ml (5fl oz) double cream

1 vanilla pod – use the seeds, which you scrape off

1/2 teaspoon finely grated orange zest

150g (5oz) caster sugar

4 leaves gelatine, soaked in cold water for 4 minutes

600ml (1 pint) sheep's yoghurt

   (cow's milk yoghurt is fine, but I prefer the sharpness given by sheep's yoghurt)

350ml (12fl oz) double cream, lightly whipped

Bring the wine, sugar, star anise and cinnamon to the boil in a pan. Lightly cut the pointed end of the tamarillos with a sharp knife to make an 'X', place them in the boiling liquid and return to the boil, then turn the heat down. Gently simmer for 5 minutes until the fruit is cooked – test by inserting a skewer. Leave to cool in the liquid.

Now for the bavarois. Have ready one large mould or six small ones. Bring the double cream, vanilla seeds, orange zest and caster sugar to the boil. Remove from the heat and add the gelatine, stirring until dissolved, then leave to go cold. (If the mixture sets before the next step, warm carefully over a pot of hot water, mixing well until it softens.) Whisk in the yoghurt, then gently whisk in the whipped cream. Pour into the mould or moulds and leave to set in the fridge for at least 4 hours.

The tamarillo is originally from South America but grows really well in the warmer climes of New Zealand. When I was a child it was known as 'tree tomato' (and kiwi fruit as 'Chinese gooseberry'). Tamarillos are very sharp-tasting, so they usually need a bit of sugar to make them edible. Here the bavarois is a perfect foil to them. This was our main dessert for Christmas 1996, served with a brandy snap (see page 188).

brandy snaps

50g (1³/4oz) golden syrup
50g (1³/4oz) caster sugar
50g (1³/4oz) unsalted butter
50g (1³/4oz) baking flour, sifted

Put the syrup, sugar and butter into a pan and boil until the mix goes a light brown colour, stirring continuously. Remove from the heat and stir in the flour, avoiding lumps. Return to the stove and cook, stirring continuously, for 2 minutes (it will colour a little in this time). Remove from the heat and leave to cool for 10 minutes.

Preheat the oven to 200°C/400°F/Gas 6. Take a teaspoonful of the brandy snap mixture in your hand and flatten it out as thin as you can, then place on a baking tray covered with non-stick baking parchment. The snaps spread quite a lot, so start with just four per tray. Place in the oven and cook for 5-8 minutes. They are ready when dark golden and quite bubbly. Remove the tray from the oven and allow to cool a bit. When they are manageable remove from the parchment and fold into cone shapes, resting them in coffee cups until they set. When cold, store in an airtight container in a cool place.

roast apricots & ginger

FOR SIX
12 apricots
2 fingers of ginger, peeled and finely julienned
150g (5oz) demerara sugar
150ml (5fl oz) water

Set the oven to 180°C/350°F/Gas 4. Mix all the ingredients in a large bowl, then gently decant into a roasting dish that's just large enough to hold it all. Bake for 20-25 minutes until the fruit is just cooked.

Served warm or cold, this is a simple, delicious and slightly tart dessert that's suitable for rounding off a rich meal. My stepmother Rose would often roast pears for a warm winter dessert, and I owe this dish to her. The ginger goes exceptionally well with the apricots. Mascarpone is the perfect creamy foil to the sharpness of the fruit and spice.

stone-fruit & raspberry vinegar salad
with shortbread

FOR SIX

200g (7oz) raspberries

150ml (5fl oz) balsamic vinegar

4 ripe peaches

4 ripe nectarines

4 ripe apricots

2 cups mixed berries

1/2 cup icing sugar

1 tub mascarpone

SHORTBREAD

250g (9oz) unsalted butter

3/4 cup caster sugar

1 1/4 cups cornflour

1 1/2 cups baking flour

1/2 cup ground almonds

Mash or purée the raspberries and mix with the vinegar, pour into a jar and leave in the fridge for a day. Then pour the mixture into a fine sieve and press all the juice through, discarding the seeds, and return to the fridge. Cut the stone fruit into halves and remove the stones. Put the fruit and mixed berries in a bowl (non-aluminium), add the icing sugar, mix gently and store in the fridge for 6 hours. An hour before serving take the fruit from the fridge and gently mix in the raspberry vinegar.

For the shortbread, set the oven to 170°C/340°F/Gas 3–4. Cream the softened butter and caster sugar. Sieve the flours and add to the butter, mixing well, then add the ground almonds and work gently into a dough. If it seems a little sticky add a little more sifted flour. Either roll the dough into logs and rest for 30 minutes before cutting into slices, or rest the dough for 30 minutes before rolling out and cutting with a biscuit cutter. Place on a baking tray you have lightly buttered or covered with baking parchment and put into the preheated oven for 20 minutes. The biscuits should go a pale gold colour. Remove from the oven and cool on the tray for 2 minutes before transferring to a rack. When cold, store in an airtight container.

Serve the fruit salad with the shortbread and mascarpone.

Make this salad first thing in the morning and serve for a last course in the evening. Peaches have a real affinity with raspberries, as peach Melba shows, but so do nectarines and apricots. Use a good-quality balsamic vinegar. The shortbread recipe is, once again, from my grandmother Molly. I would serve the salad with whipped cream.

roast quince
with maple syrup

The quince is a wonderful fruit. It looks beautiful, the aroma when ripe is exquisite and it imparts its subtle flavour to a range of delicacies, like cottognata (quince cheese), pickled quince, ice creams and spicy relishes. The recipe given here is great served with whipped cream and a crunchy biscuit or with a rich goat's or sheep's cheese such as Golden Cross or Little Rydings. Only a few varieties of quince are grown commercially and most turn a deep red when cooked. One variety goes a rich golden colour, and this was used for the accompanying photograph.

FOR FOUR

4 quinces

1 cinnamon stick

2 bay leaves

1 lemon (remove the rind with a potato peeler and juice it)

150ml (5fl oz) maple syrup or honey

350g (12oz) caster sugar

Set the oven to 180°C/350°F/Gas 4. Wash the quinces under warm water and rub off the fine fur if they are the variety that have it. Cut them in half and remove the hard core. Put the cinnamon, bay leaves, lemon peel and juice into a ceramic roasting dish and sit the quinces on top, cut side up. Drizzle the maple syrup over them and sprinkle with the sugar, then pour in enough hot water to come halfway up the fruit. Cover with foil, making a tight seal, and bake for 2 hours. Check after 90 minutes: the quinces are cooked when a skewer goes through them easily. Let them cool in their syrup as they are very fragile. In the fridge they will keep in the syrup for a few weeks.

spicy poached pear
with mascarpone

'Spicy' is a word that appears frequently in my menus, but some are surprised to see it in the desserts section. In my youth, if there was a cookbook that every New Zealand mother bought for her child when they left home, it was the *Edmonds Cookbook*; no good household was without its copy. This contained something described enticingly as 'Araby Spice Cake', which was not complicated but very tasty. Then there were the 'Easter Spice Biscuits' – which my sisters and I would knock up in batches. I remember, too, making 'Tomato Soup Cake' from the *New Zealand Radio Times Cookbook*. This was my step-brother Dean's favourite. It had a lot of ground spice in it and, as is the way when you make something often, you experiment – and one day I added Tabasco. It worked a treat, so I wondered what other desserts this might improve. The following recipe is a direct result of those experiments and is also delicious if you add extra chilli and serve as a first course with gorgonzola and salad greens.

6 firm cooking pears, such as Comice
1 red chilli
1 bay leaf
6cm (2–3in) rosemary sprig
2 star anise
2 cinnamon sticks
3 cloves
Zest and juice of 2 lemons
400g (14oz) unrefined caster sugar
2 tablespoons honey
600ml (1 pint) good red wine (but no call for a vintage!)
300g (11oz) mascarpone

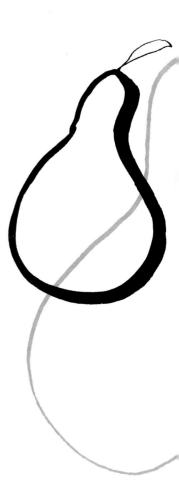

Cut the chilli in half and add, with all the other ingredients except the pears and the mascarpone, to a saucepan just large enough to hold the pears. Bring to the boil and simmer for 5 minutes. Peel the pears and add to the liquid, adding water if necessary to cover the fruit. Cut out a circle of greaseproof paper slightly larger than the pot and place over the pears, pressing down so it comes into contact with the liquid. This will ensure even cooking – more so than a lid as it is close to the fruit. Bring back to the boil and continue to simmer for at least 30 minutes (the time will depend on the size and ripeness of the fruit). The pears are cooked when a thin skewer is easily inserted through the centre.

When done, remove the pears with a slotted spoon and place gently in a bowl. Turn the heat up and boil the liquid in the pan until reduced by half, then strain over the fruit. The pears can be served either hot or chilled with a generous dollop of mascarpone.

date crème brûlée

This is a slight twist on the traditional French dessert, and one which I prefer. For another mouthwatering variation replace the dates with 6–8 raspberries or 2 teaspoons of fresh passionfruit pulp per ramekin to make a sensational summer dessert.

FOR SIX

900ml (1¹/₂ pints) double cream
9 egg yolks
200g (7oz) caster sugar
¹/₂ vanilla pod
9 medjool dates, pitted and cut into quarters[halves?]
Demerara sugar for glazing

You'll need six 250ml (9fl oz) ceramic ramekins and a roasting tray as deep as the ramekins and just big enough to hold them. Bring a kettle full of water to the boil and turn the oven to 170°C/340°F/Gas 3-4.

Scrape the seeds from the vanilla pod and put in a pot with the cream and half the sugar (the stripped pod can go in a vanilla essence bottle to spice it up). Bring the cream to the boil and keep warm. Lightly whisk the yolks and the remaining sugar for 30 seconds, then gently pour in the hot cream and mix well.

Put three pieces of date in the bottom of each ramekin, arrange the ramekins in the roasting tray and divide the mixture between them. Place in the oven and gently pour in the hot water until it comes two-thirds of the way up the sides of the ramekins. Bake for 30 minutes. Test by inserting a skewer – it should come out clean; if it doesn't, return to the oven till done. Make sure the brûlées don't brown on the top or bubble up: if they do, turn the oven down a bit. When cooked, remove from the oven and stand for an hour before placing in the fridge for at least 6 hours to firm up.

Just before serving sprinkle 2 teaspoons of demerara sugar in a thin layer over the top of each one and place under a grill on high heat. Keep an eye on them as the sugar will turn to burnt toffee quite quickly – the aim is to give the brûlées a filmy topping of caramel. Allow to cool for a minute before serving as the toffee will burn your lips.

bread

making bread

Making bread is both easy and very rewarding. It's something we always do at the London Sugar Club, and it was the same in Wellington. So if you visit you're likely to find the bread different each time as we tend to let our fancy fly! I remember one loaf we made in New Zealand from fresh, chopped peaches and some over-dry blue cheese lying forgotten at the back of the fridge. It was delicious.

The guidelines I give to anyone making bread for the first time are as follows. Leavened bread is made from yeast, water and flour, and that's all. There are many types of flour, and the one usually preferred for bread is called 'strong' flour, which means it has a high content of gluten, the protein that gives flour its strength. I have made quite fine bread using flour, though, so don't despair if you can't get the strong stuff.

Yeast is a living organism, and the things that kill it are salt and water that's too hot. You could make a 100kg (220lb) loaf with just 1 teaspoon of yeast, but it would be a few days before it rose properly. The same loaf could be made using 20kg (45lb) of yeast, but this would taste too yeasty. So the balance of yeast to the other ingredients must be judged correctly to produce a satisfying loaf or bun. It's also important that the liquid you add to the yeast is neither too hot nor too cold. Too cold and it will take forever for the yeast to become active. Too hot and it dies. Add salt when you add the flour or this too may kill the yeast. After making, the dough should be covered with clingfilm and left in a warm place to prove. If left uncovered it will form a skin. If it is left in too hot or too cold a place it just won't work that well.

So here are four yeast-based bread recipes, all quite different. I've used white flour in each of them – at the Sugar Club we use only organic unbleached white flour and throw in polenta, rye flour or wholemeal to provide variety. Remember, though, that you can easily knock up a batch of great-tasting bread using nothing more than warm water, yeast and flour.

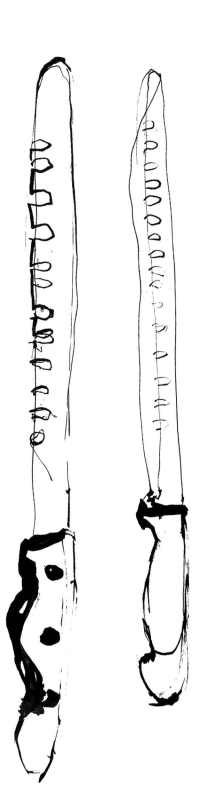

walnut
bread

200ml (7fl oz) milk, at body temperature
1 egg
3 egg yolks
30g (1oz) brown sugar
20g (³/₄oz) fresh yeast
300g (11oz) plain flour
250g (9oz) walnut halves
¹/₂ teaspoon salt
A glaze made from 1 egg white mixed with 2 tablespoons milk

Set the oven at 200°C/400°F/Gas 6 and line a 25cm (10in) loaf tin with non-stick baking paper. In a large bowl mix the milk, eggs, sugar and yeast and leave to sit in a warm place for 10 minutes. Sieve the flour into the bowl and mix well. Add the nuts (and fruit if you're using it – see below) and, last, the salt. Pour into the loaf tin and put in a warm place to prove (30-45 minutes).

When the mixture has risen to within 1cm (¹/₂in) of the top of the tin, gently brush with the glaze and place in the oven. After 5 minutes turn the temperature down to 170°C/340°F/Gas 3–4 and cook for a further 25 minutes. Keep an eye on the loaf and make sure the top doesn't burn; if it does overbrown, lay a piece of tin foil loosely over it. The bread is cooked when a knife inserted into it comes out clean or when a gentle tap gives a hollow sound. Allow it to cool in the tin for 5 minutes and then tip on to a cake rack, remove the paper and leave to cool completely.

This is my favourite thing to have with cheese and, when it gets a bit stale, it's just as appetising toasted with goat's cheese on top. The mix resembles a thick batter when made, so don't worry that it doesn't look like a bread dough. For variation, add sultanas that have been soaked in warm water for half an hour and other dried fruit and nuts.

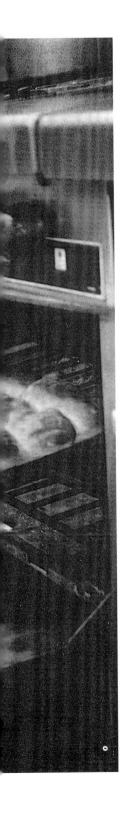

sweet
potato
bread

This couldn't be simpler. You'll end up with an irresistible orange-coloured bread that's slightly sweet and rich. If you can't get sweet potatoes, use ordinary potatoes or pumpkin: the result will of course differ but will be just as tempting. (For photograph see page 194.)

FOR TWO 750G (1LB 11OZ) LOAVES

500g (18oz) sweet potato, peeled and cut into 2cm (^{3}/$_4$in) dice
400ml (14fl oz) milk
20g (^{3}/$_4$oz) fresh yeast (or 3 teaspoons dried yeast), dissolved in
*   50ml (1^3/$_4$fl oz) warm water*
800g (1^3/$_4$lb) strong flour
1 teaspoon sea salt

Put the sweet potato and milk in a pot, put the lid on and bring to the boil. Cook until the potato is done, then remove the lid and boil until the milk has reduced by half. Put the milk and cooked sweet potato into a bowl and allow to cool until lukewarm. Add the yeast dissolved in the water and mix well, then add all the flour and the salt and knead for 5 minutes (you can do this by hand or in a food mixer). The dough should be moist but not sticky, so if it needs a bit more flour just add some. Leave it in a warm place to double in size – this should take about an hour. When risen, punch the dough down with your fist and divide it into two lumps of equal size. Roll each piece into a long sausage shape and place on baking parchment on a baking tray. Again, leave in a warm place to double in size.

Turn the oven to 180°C/350°F/Gas 4. Brush the dough with warm water and sprinkle with a little coarse sea salt, then place in the top half of the oven. After 20 minutes test by tapping a loaf on the bottom – it should sound hollow. If it doesn't, cook until it does; this should take no more than another 10 minutes. Remove the bread from the oven and cool on a cake rack.

spicy
plantain bread

FOR TWO 750G (1LB 11OZ) LOAVES

20g (³/₄oz) fresh yeast (or 3 teaspoons dried yeast)

500ml (18 fl oz) warm water

200g (7oz) polenta grains, sieved to remove lumps

2 tablespoons molasses sugar

180g (6¹/₂oz) grated peeled plantain (use one that's just starting to go brown)

750g (1lb 11oz) strong flour

4 teaspoons hot paprika (smoked paprika is also excellent in this recipe)

50ml (1³/₄fl oz) olive oil

30ml (1fl oz) Asian fish sauce

Dissolve the yeast in the water, then mix in the polenta, sugar and plantain. Add a quarter of the flour, mix well, and leave in a warm place for 10 minutes. Add the paprika, olive oil, fish sauce and half the remaining flour and mix well together, then add the rest of the flour and knead well for 5 minutes, either by hand or in a mixer with a dough hook attached. The dough should be moist but not sticky: if it's too wet add more flour; if it's too dry add a little warm water and knead again for a minute. Cover the bowl with clingfilm and leave in a warm place for 40 minutes.

Turn the oven to 190°C/375°F/Gas 5 and sprinkle a baking tray with an extra half cup of polenta grains. Divide the dough into two and gently stretch each piece into a rather rugged-looking log 30cm (12in) long. Put them on the tray, brush with warm water and then sprinkle each loaf with 3 tablespoons of polenta. Leave in a warm place for 20 minutes before placing in the top half of the oven and cooking for 35 minutes. Test that the bread is cooked by tapping the bottom of a loaf – it should sound hollow; cook for a little longer if it isn't ready. Remove the bread from the oven and brush each loaf with an extra 30ml (1fl oz) olive oil, place on a cake rack and leave to cool a little before eating – if you can resist that long.

This bread, like a lot of my food, has a combination of tastes. The sweetness of the plantain and molasses sugar is offset by the heat from the paprika (which, if you want, can be replaced by finely chopped fresh green chilli). This sort of flavour-packed bread is best served with something simple like grilled fish or roast chicken, though as it's a favourite of mine I'll eat it with anything. It's delicious with a very ripe goat's cheese on top for a late supper.

grilled yoghurt & nigella seed
flat breads

This bread takes its inspiration from the roti, the tasty griddle-cooked flat bread from India. I use yeast in this recipe, though it's not usual to do so. Nigella seeds, called 'kalonji' in parts of India, are savoury/aromatic black seeds that give the bread much of its character. If you can't find them, cumin, crushed coriander or fennel seeds work almost as well. I like to eat flat bread with saucy dishes, like curry, but they're also great on a picnic for rolling stuffings in – a little like pitta breads.

FOR 16 FLAT BREADS

2 teaspoons nigella seeds
125ml (4^1/$_2$fl oz) cold water
250g (9oz) goat's milk yoghurt (or use cow's yoghurt)
10g (^{1}/$_3$oz) fresh yeast (or 2 teaspoons dried yeast)
200g (7oz) strong flour
^{1}/$_4$ cup finely sliced spring onions
1 teaspoon salt
30ml (1fl oz) sunflower oil
350g (12oz) bread flour

Put the nigella seeds and water into a small pot and boil for 1 minute. Cool a little. Tip into a large bowl and whisk in the yoghurt, then the yeast, then the first amount of flour. Cover the bowl with clingfilm and sit in a warm place for 20 minutes. Stir in the spring onions, salt and oil, and then tip in the second lot of flour. Mix with your hands until the flour is taken up, and knead for 1 minute. The dough should be moist but not sticky. Cover with clingfilm and rest for 15 minutes in a warm place.

Heat up a skillet, heavy pan or griddle to a medium-high heat. Divide the dough into four equal pieces, then each piece into four again. Sprinkle them all lightly with a little extra flour and roll out, one piece at a time, to a thickness of 3mm (^{1}/$_8$in) with a rolling pin. I'm quite happy if they come out funny shapes, but roll them into perfect circles if you think that's more pleasing.

Now begins the production line. As soon as the first has been rolled, brush it lightly with sunflower oil and lay it on the grill, oiled side down. While it's cooking – they need about 1 minute on each side – roll the next, and so on. Be careful they don't burn, but make sure they are cooked on the inside. Practice will tell you when they're ready. Stack them on a plate as they come out of the grill and cover with a cloth to keep them warm if they're going to the table. You can also fry these breads – roll them out as normal, but instead of grilling shallow-fry or deep-fry them at 180°C/350°F for a minute on each side.

glossary

Aïoli In its basic form, a garlic mayonnaise.

Asian fish sauce Made from fermented anchovies and sour and salty in taste, this is an essential ingredient in Southeast Asian cooking. Available from Thai food stores and larger supermarkets.

Arame An edible seaweed that contributes texture and taste to a dish. Available from health food shops and Japanese food stores.

Balsamic vinegar An aged vinegar, slightly sweet and dark brown, originating from Modena, Italy. It comes in various grades, the very best selling for hundreds of pounds a bottle. Very cheap balsamics may be inferior, so beware! Sold in all supermarkets and delicatessens.

Blanching A cooking method where the raw ingredients are plunged briefly into boiling water and then drained or refreshed in cold water.

Bavarois A creamy dessert, often set with gelatine.

Bisque A thin seafood broth, usually based on shellfish and often with lots of fish and crustaceans added.

Blachan Also called 'kapi' or shrimp paste, this comes as a compressed block of rather foul-smelling dried shrimps. When toasted or fried it imparts a delicious underlying taste to curries and soups.

Bok choy 'Choy' translates more or less as 'cabbage', and many types are available – bok choy, pak choy, choy sum, choy sim, etc. Rather similar, it's their shape, texture and colour that affect a dish more than their flavour, which is crisp and fresh.

Bruschetta A piece of bread, usually sourdough, rubbed with garlic and drizzled with olive oil. Bruschetta is served with soup, with a topping as a snack or as an accompaniment to a dish. Use a slightly stale, farmhouse-style bread. Fresh white bread won't do.

Buckwheat Buckwheat flour is made from the dried ground seeds of the buckwheat plant and is a principal ingredient of soba noodles (qv) and Russian-style *blini* (pancakes). Whole buckwheat can also be steamed or boiled briefly to use as a starch like couscous.

Canapé A small, tasty treat, usually savoury, that can be held with two fingers.

Ceviche Also known as 'seviche'. This dish – raw fillets of fish 'cooked' in citrus juice – is eaten throughout South America and is popular in Spain and Portugal. The technique of cooking in citrus juice is also known in the islands of the South Pacific.

Chard Also called 'Swiss chard' or 'blett' in Europe and 'silverbeet' in New Zealand and Australia. Most readily available in the green and white form, but a fabulous deep red and green variety is becoming popular.

Chorizo A Spanish sausage with as many varieties as Italian salami. There are cooking varieties and those that can be eaten raw. 'Pata negra', produced from 'black foot' pigs fed on acorns, is said to be the best. Chorizo usually has a high content of smoked paprika, giving it a distinctive red colour.

Choy sum See **Bok choy**.

Cornichons Tiny gherkins, usually pickled.

Crostini Thin slices of bread drizzled with olive oil and baked until crisp and golden. In many ways they are the Italian croûton. Often served with a tasty topping as a canapé (qv) or in soups or salads.

Daikon An important vegetable in Japanese, Korean and Chinese cooking, also known as 'mooli', 'white radish', 'winter radish' or 'giant white radish'. It looks like a large white parsnip, but its smooth white skin is closer to our radishes – the taste is similar though not as intense. Used in Japan shredded as an accompaniment to *sashimi*, in Korea as a pickle, and carved into exotic shapes as a garnish.

Dariole mould A mould, usually metal and large enough for a single portion, in which jellies and various bavarois are set or small steamed puddings can be made.

Flour, plain Flour with a low gluten content used to make biscuits and cakes.

Flour, strong Flour with a high gluten content and so ideal for breadmaking. Gluten, a protein, gives a loaf or bun its strength and makes it good and chewy to eat.

Galangal A member of the ginger family but one that is more fibrous and with a very pungent smell and taste. The best quality are pale in colour and firm. Fresh is best, but use the dried, powdered form if necessary. Add sparingly to soups and curries. Available in most Southeast Asian food stores and some larger supermarkets.

Ganache A mixture of melted chocolate and cream or butter; used to make chocolate truffles or to cover cakes.

Harissa A paste, based on hot red chilli peppers, olive oil and garlic, which is used in North African countries as a flavouring, sauce or condiment. Sold in tins in Middle Eastern stores.

Hijiki A dried seaweed, with a great texture, that looks like an intensely black, curly spaghetti. It must always be soaked before using. Available from health food shops and Japanese food stores.

Kombu A giant sea kelp which is one of the most important ingredients of Japanese cuisine. It is a principal flavouring in *dashi*, the stock base of many Japanese savoury dishes. Dried and packaged in several forms, it's available from health food shops and Japanese food stores.

Laksa A soup from Malaysia and Singapore containing noodles and fish or meat or a combination of both. Often spicy, and sometimes made with coconut milk.

Lemon grass The stem of a wild, grass-like bush that grows throughout Southeast Asia. Available from Asian stores and larger supermarkets.

Mascarpone An Italian soft cream cheese made from cow's milk. It is like thick double cream in texture and flavour and is eaten fresh as a dessert, usually with fruit. Also an essential ingredient in *tiramisu*.

Medjool An intensely rich, soft fresh date.

Mirin A Japanese sweetened rice wine, never drunk but used for sauces, marinades, etc.

Miso A semi-solid paste made from fermented soy beans and sold in red, white and brown varieties. White is the most subtle. Some are made with added barley or rice. Keep in the fridge in an airtight container. Available from health food shops and Japanese food stores.

Mooli See *Daikon*.

Mouli or **mouli-légumes** An implement for cutting, grating and puréeing vegetables. Not to be confused with mooli.

Nigella The tiny, jet-black seeds of a wild onion; often known as black cumin and, in India, as 'kalonji'.

Nori A highly nutritious seaweed which is harvested in Far Eastern waters and then dried and compressed into flat sheets. Mainly used as a wrapping for *sushi*, it can also be shredded and added to soups or used as a condiment. Its flavour is increased if it is toasted before using. Nori is known as 'laver' in Wales, where it is made into laverbread. Available from health food shops and Japanese food stores.

Pak choy See **Bok choy**.

Pancetta Italian streaky bacon that comes raw or smoked ('affumicata') and is used in dishes like spaghetti carbonara. The flavour is magnificent. The best Spanish variety, which is as good as the best Italian, is spelt 'panceta' and has smoked paprika rubbed into it.

Pecorino An Italian ewe's milk cheese which can be young and soft or aged and hard. The best of the latter are Pecorino Sardo (from Sardinia) and Pecorino Romano.

Pesto Almost literally a 'paste', pesto is a ground or puréed cold sauce. Originally made from garlic, pecorino (*qv*), basil, olive oil and pine nuts, pesto can be made from a huge variety of ingredients.

Polenta Grains consisting of ground dried corn or maize kernels. Both the grains and the cooked dish are referred to as 'polenta'.

Radicchio A bitter red lettuce with quite thick leaves; often braised, grilled or eaten raw.

Render A method of removing fat from meat by heating it on a moderate heat until the fat melts away.

Ramekin A small ceramic dish used to bake food in.

Salamander The term used in industry and restaurant kitchens to refer to an overhead grill.

Sauté To shallow-fry in fat or oil in a frying pan or sauté pan, usually colouring the food at the same time.

Shallot A small member of the onion family with a pale brown skin.

Shiitake Japanese mushrooms cultivated on types of oak tree and sold fresh or dried. Known as 'winter mushrooms' in China. Available from Japanese (and good Chinese) food stores and larger supermarkets. Soak the dried variety in cold water for 3 hours before use.

Skillet A flat item of cooking equipment, usually made of cast iron, that is seated directly on a heat source.

Soba noodles Japanese noodles of medium thickness and square in shape. They are made from a mixture of buckwheat (qv) and wheat flours and come fresh or dried. I prefer those with a 40% buckwheat content. Available from Japanese food stores and health food shops.

Star anise The star-shaped seed pod of a plant that is widely grown in southern China. A spice with a heady aroma, it is used in many Chinese and some Southeast Asian dishes. It's great with rich foods like pork, duck and *chocolate*! It comes whole or ground. Now stocked by larger supermarkets and almost any Asian food store.

Suprêmes Chicken suprêmes is just chef-speak for chicken breasts.

Sushi ginger Lightly sweet-pickled fine slices of root ginger, used to accompany *sashimi* and *sushi*. Available from Japanese food stores.

Sushi mat Looking like a bamboo place mat, a sushi mat consists of a series of fine bamboo slivers bound with twine, allowing it to be rolled into a cylinder. Available from Japanese food stores.

Tabasco A very spicy red sauce made from matured peppers. Used in Bloody Marys and in cooking when you need to give something a big kick.

Tamari A wheat-free soy sauce with a sharper taste and stronger aroma than *shoyu*, another premium Japanese soy sauce. Most soy sauces have wheat added to thicken them, but I find this unnecessary. In this book I have specified tamari, but any good-quality soy sauce works well in all the recipes. Available from Japanese food stores, health food shops and larger supermarkets.

Tamarind A magnificent tree that grows in Southeast Asia and India. The seed pods resemble medium-sized broad bean pods, and it is from these that the tamarind paste is extracted. Sharp and sour in taste, it is essential in most Asian cuisines. It can be bought in paste form (which is the easiest to use) or, occasionally, compressed. The compressed form has to be boiled up in water, after which you strain out all the fibrous matter.

Tapenade A paste made from black olives, anchovies and capers (green olives may also be used, albeit non-traditionally). It is used as a dip, spread, filling or sauce. The name comes from the Provençal word for caper.

Wasabi A Japanese plant, grown in flowing water, whose roots yield one of the hottest spices. The bright green flesh of the roots is grated and used fresh or is dried and ground to be made into a paste, like English mustard powder; this 'green horseradish' paste is used as an accompaniment to *sushi* and *sashimi*.

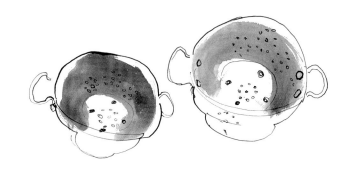

index